T0299742

Implementing Enterprise Cyber Security with Open-Source Software and Standard Architecture
Volume II

RIVER PUBLISHERS SERIES IN DIGITAL SECURITY AND FORENSICS

Series Editors:

ANAND R. PRASAD
Deloitte Tohmatsu Cyber LLC, Japan

R. CHANDRAMOULI
Stevens Institute of Technology, USA

ABDERRAHIM BENSLIMANE
University of Avignon, France

PETER LANGENDÖRFER
IHP, Germany

The "River Publishers Series in Security and Digital Forensics" is a series of comprehensive academic and professional books which focus on the theory and applications of Cyber Security, including Data Security, Mobile and Network Security, Cryptography and Digital Forensics. Topics in Prevention and Threat Management are also included in the scope of the book series, as are general business Standards in this domain.

Books published in the series include research monographs, edited volumes, handbooks and textbooks. The books provide professionals, researchers, educators, and advanced students in the field with an invaluable insight into the latest research and developments.

Topics covered in the series include:-

- Blockchain for secure transactions
- Cryptography
- Cyber Security
- Data and App Security
- Digital Forensics
- Hardware Security
- IoT Security
- Mobile Security
- Network Security
- Privacy
- Software Security
- Standardization
- Threat Management

For a list of other books in this series, visit www.riverpublishers.com

Implementing Enterprise Cyber Security with Open-Source Software and Standard Architecture

Volume II

Editors

Anand Handa

C3i Center, Indian Institute of Technology, Kanpur, India

Rohit Negi

C3i Center, Indian Institute of Technology, Kanpur, India

S. Venkatesan

Indian Institute of Information Technology, Allahabad, India

Sandeep K. Shukla

C3i Center, Indian Institute of Technology, Kanpur, India

River Publishers

Routledge
Taylor & Francis Group
NEW YORK AND LONDON

Published 2023 by River Publishers
River Publishers
Alsbjergvej 10, 9260 Gistrup, Denmark
www.riverpublishers.com

Distributed exclusively by Routledge
605 Third Avenue, New York, NY 10017, USA
4 Park Square, Milton Park, Abingdon, Oxon OX14 4RN

Implementing Enterprise Cyber Security with Open-Source Software and Standard Architecture / by Anand Handa, Rohit Negi, S. Venkatesan, Sandeep K. Shukla.

Routledge is an imprint of the Taylor & Francis Group, an informa business

ISBN 978-87-7022-795-7 (print)
ISBN 978-87-7022-962-3 (paperback)
ISBN 978-10-0092-238-7 (online)
ISBN 978-1-003-42613-4 (ebook master)

While every effort is made to provide dependable information, the publisher, authors, and editors cannot be held responsible for any errors or omissions.

Contents

Preface

The Interdisciplinary Center for Cyber Security and Cyber Defense of Critical Infrastructures (C3i Center) at Indian Institute of Technology Kanpur, India, in association with Talent Sprint, organized the Cohort-3 of a six-month long advanced cybersecurity training program from Feb 2021 to August 2021. Through this program, we trained approximately 50 IT professionals from various domains and MNCs in India and abroad. The rigorous training program offered weekly live training sessions, homework, quizzes, projects, and a final capstone team project. The participants were motivated to learn various new cybersecurity domains and apply the lessons learned to develop cybersecurity solutions or architectures from open-source tools. The course had IT professionals with substantial experience from different industry segments. Being experienced, most of them accepted the challenge of developing tools by learning and utilizing, modifying, and integrating available security solutions from the open-source domain.

This book has nine chapters describing the projects that participants took as capstone projects. The book explains in detail the tools and methodologies used in developing them using open-source tooling to obtain common cyber defense, malware analysis tools, automation for self-penetration testing, and vulnerability assessment. The book has four categories of chapters that describe the methodologies of developing tools for – malware analysis using machine learning, deploying honeypots, Intrusion Detection Systems (IDS), and web application security.

We thank all the contributors to this book. We must thank Mrs. Debjani Mukherjee of Talent Sprint for her contribution to proofreading all the chapters. We also thank all the professionals participating in the Talent Sprint advanced certification program offered by C3i Center at IIT Kanpur. We especially thank all those who readily agreed to contribute chapters for the book despite their busy professional schedule. We thank Mr. Rohit Agarwal, Mr. U. Prasad, and Dr.Santanu Paul from Talent Sprint for enabling the course continuously for six months throughout all weekends. Finally, we thank Prof. Manindra Agrawal for supporting the program, the IIT Kanpur Center for Continuing Education

(CCE) staff, and the head of CCE for their cooperation in offering this training program. We also thank Mr. Nitesh Kumar, Mr. Subhasis Mukhopadhyay, and other member staff at the C3i center for their sustained help from the background during the training period and beyond.

We hope this book will be helpful to many, and we plan to develop similar books for future cohorts of trainees joining through this program.

List of Figures

List of Tables

List of Contributors

Annamalai, Balaji, *Independent Researcher;*
E-mail: Balajiannamalai@hotmail.com

Chawla, Kunal, *Cyber Security Consultant;*
E-mail: kunaalchawla@yahoo.com

Goswami, Yathartha, *Under Graduate Researcher, C3i Center, Indian Institute of Technology, Kanpur, India*

Handa, Anand, *Post-Doctoral Fellow, C3i Center, Indian Institute of Technology, Kanpur, India; E-mail: anand@c3ihub.iitk.ac.in*

Jagadeesan, Senthil, *Independent Researcher;*
E-mail: jj_senthil75@yahoo.co.in

Kashamshetty, Shankar, *Independent Researcher;*
E-mail: kshankar14@gmail.com

Khedar, Ranveer, *Independent Researcher;*
E-mail: ranveerkhedar@gmail.com

Kumar, Nitesh, *Senior Project Engineer, C3i Center, Indian Institute of Technology, Kanpur, India; E-mail: niteshkr@cse.iitk.ac.in*

Majumdar, Partha, *E-mail: partha.majumdar@hotmail.com*

Mishra, Om, *Independent Researcher, E-mail: mishra.om@live.com*

Mishra, Om Prakash, *Chief Solutions Architect, Canum Infotech, India;*
E-mail: om@canuminfotech.com

Negi, Rohit, *Senior Vice President (R&D) and Lead Engineer, C3i Center, Indian Institute of Technology, Kanpur, India; E-mail: rohit@cse.iitk.ac.in*

Patlan, Atharv Singh, *Under Graduate Researcher, C3i Center, Indian Institute of Technology, Kanpur, India; E-mail: atharvsp@iitk.ac.in*

Sahni, Neelakshi, *HOD-Computer Dept, CCA School, Sector 4, Gurgaon, Haryana, India; E-mail: neela.creative2020@gmail.com*

Sarkar, Ria, *Independent Researcher; E-mail: riasarkar1999@gmail.com*

Seludkar, Kalpesh, *Independent Researcher;*
E-mail: kseludkar13@gmail.com

Shukla, Sandeep K., *Professor & Program Director, C3i Center, Indian Institute of Technology, Kanpur, India; E-mail: sandeeps@cse.iitk.ac.in*

Singh, Amardeep, *Program Head Cyber Security, eclerx services Ltd., India; E-mail: amardeepsg@gmail.com*

Soni, Vishal, *Founder, SVS Datalytics, Udaipur, Rajasthan, India; E-mail: soni.vs@gmail.com*

Srivastava, Purushartha, *Independent Researcher, E-mail: purusharthasrivastava1993@gmail.com*

Sumbria, Sanjeev Kumar, *Independent Researcher; E-mail: sam288037@gmail.com*

Suvarna, Sheetal A., *Independent Researcher, E-mail: sheetalasu-varna@gmail.com*

Tambe, Som Vishwas, *Under Graduate Researcher, C3i Center, Indian Institute of Technology, Kanpur, India; E-mail: somvt@iitk.ac.in*

Tripathi, Shyava, *Independent Researcher; E-mail: shyava.tripathi8@gmail.com*

Yadav, Ashish Ranjan, *Independent Researcher; E-mail: aryadav@cse.iitk.ac.in*

List of Abbreviations

AIDE	Advanced intrusion detection environment
API	Application programming interface
AUC	Area under the ROC curve
BMP	Bitmap image file
BYOD	Bring your own device
CDB	Constant database
CFHC	Coastal family health center
CIS	Center for internet security
CNN	Convolutional neural network
CPE	Common platform enumeration
CSV	Comma-separated values
CUPS	Common unix print system
CVE	Common vulnerabilities and exposures
CWE	Common weakness enumeration
DCCP	Datagram congestion control protocol
DGCNN	Dynamic graph convolutional neural network
DHCP	Dynamic host configuration protocol
DNS	Domain name system
ELF	Executable and linkable format
ELK	Elasticsearch, Logstash, and Kibana
EXIF	Exchangeable image file format
FIM	File integrity monitoring
FPR	False-positive rate
GHDB	Google hacking database
HIDS	Host intrusion detection system
IDS	Intrusion detection system
IOC	Indicator of compromise
IoT	Internet of things
JFIF	JPEG file interchange format

JPEG	Joint photographic experts group
LAMP	Linux, Apache, MySQL, and PHP
LDAP	Lightweight directory access protocol
LSTM	Long short-term memory
MA	Modified availability
MAC	Modified attack complexity
MAV	Modified attack vector
MC	Modified confidentiality
MHN	Modern honeypot network
MI	Modified integrity
MPR	Modified privileges required
MS	Modified scope
MSME	Micro, small and medium enterprises
MUI	Modified user interaction
NFS	Network file system
NMHC	Northwestern memorial healthCare
NVD	National vulnerability database
OEM	Original equipment manufacturer
OSINT	Open-source intelligence
OSS	Open-source software
OWASP	Open web application security project
PAM	Pluggable authentication modules
PPA	Personal package archive
RBAC	Role-based access control
RDBMS	Relational database management system
RDS	Reliable datagram sockets
ROC	Receiver operating characteristic
RPC	Remote procedure call
SaaS	Software as a service
SCA	Security configuration assessment
SCP	Secure copy protocol
SCTP	Stream control transmission protocol
SFTP	Secure file transfer protocol
SIEM	Security incident and event management
SMOTE	Synthetic minority oversampling technique
SNMP	Simple network management protocol

SSH	Secure shell
TF-IDF	Term frequency-inverse document frequency
TIPC	Transparent inter-process communication
Tmpfs	Temporary file storage filesystem
TNR	True negative rate
TPR	True positive rate
VLAN	Virtual local area network

Part I

Web Application Security

1

OWASP G0rKing – Exploiting the Hidden Aspects of Google's Search Capabilities

Vishal Soni[1] and Neelakshi Sahni[2]

[1]SVS Datalytics, Udaipur, Rajasthan, India
[2]CCA School, Sector 4, Gurgaon, Haryana, India
E-mail: soni.vs@gmail.com; neela.creative2020@gmail.com

Abstract

Cyberattacks have become a common norm, and therefore it has become extremely important for organizations to keep their digital assets secured. Organizations may often unknowingly expose their critical resources exposed on the Internet without even knowing about it. While hosting and services online, some sensitive information like user credentials or configuration files may be left unattended, leading to crawling and indexing by search engines. Such ignorance may lead to exposure to critical information. Penetration testing allows organizations to test their online assets, such as websites, hosted applications, content, databases, etc., across the networks against any kind of internal or external threats. This also allows organizations to proactively identify and secure any loopholes that may lead to unauthorized access. This chapter will address the design and implementation of such a utility tool, that can help organizations proactively audit and protect their digital assets by performing penetration testing. With the help of such a tool, organizations can perform vulnerability scans and penetration testing at a regular internal level to identify any known vulnerabilities and ensure that their digital assets are secured.

1.1 Introduction

Conducting regular penetration testing can help organizations in identifying the exposed resources and securing such vulnerabilities. Pentesters can carry out reconnaissance or open-source intelligence (OSINT) gathering to proactively find such exposures before any other outsider does. OWASP G0rKing (formerly called Project SaUR0N [56]) is an open-source tool that can automate reconnaissance by using the Google dork search process. The objective of G0rKing is to automate the Google dorking process and offers a way to save the results in a text file for easy reference and reporting purposes. It can also automate the process of checking one URL against multiple dorks and can hence provide a quick health check. This fully functional command-line-based tool is available for users to perform reconnaissance tests for any website. It can be customized and fine-tuned to search for any specific type of vulnerability or resources. It can be used to search for any already indexed and publicly available information across the Internet. It also offers a way to save the results in a text file for easy reference and reporting purposes. Featured highlight: The official OWASP leadership team has agreed to take this project into their umbrella, and they have included it into their repository [62].

1.2 Literature Survey

1.2.1 What is google dorking?

When searching, you often collect as much information as possible about a topic. Advanced search techniques can help to find answers to the questions (search). For example, you are searching for a company's tax return, information that may not appear on their websites or show up when you do a regular web search.

Google dorking is a technique used by various organizations, security auditors, news agencies, investigating agencies, tech-savvy criminals to question search engines to find hidden information that might be available on public websites. As this technique is used by various search engines, we can simply refer to dorking. Dorking helps searchers to reveal the results that are not visible in the normal regular search. It helps searchers to dive deeper and refine their searches in web pages and online documents. Dorking does not need any technicality, it works using a few search techniques and using them across several search engines.

Google dorking requires a system with an Internet connection, appropriate use of search function, with appropriate use of operators which will help you to deepen your search results. To google dork, dorkers require perseverance, vision, diligence, and success is the end result.

1.2.2 A brief history of dorking

Johnny Long, aka j0hnnyhax [49], was a pioneer of dorking. He first posted his definition of the newly coined term, GoogleDork, in 2002. Since then, its meaning has evolved to include other usages.

In a 2011 interview, Johnny Long said: "*In the years I have spent as a professional hacker, I have learned that the simplest approach is usually the best. As hackers, we tend to get down into the weeds, focusing on technology, not realising there may be non-technical methods at our disposal that work as well or better than their high-tech counterparts. I always kept an eye out for the simplest solution to advanced challenges.*"

To dork or not to dork: Using the full potential of the search engines, dorking can uncover information on websites as well as various threats that they could hold. This could even include a password-protected file or folder that was to be hidden but now is vulnerable. Dorking expands the search horizon and can be an eye-opener to the information that was of general use but is not available and now could be visible through search engines.

Be Aware! If the Techie's are using Google dorking as a searching method, then few words of caution before they start!

1. Certain legal issues which could be taken care of which are involved while viewing pages and files, even if they are publicly available on the server.
2. Though search on various search engines is free but at some stage downloading some web pages or files can be prosecuted, especially under the Computer Fraud and Abuse Act in the USA.
3. The search queries are scrutinized and saved by search providers and the government, but a word of caution! These can be used against you in the future as they are very well docketed.

1.3 Purpose

G0rKing was created as the Capstone Project delivery for the completion of the COHORT-3 course and certification. COHORT-3 was offered through the joint initiative of C3i Center (IIT, Kanpur) and Talentsprint in the year 2021. The course title is "Advanced Certification Program in Cyber Security and Cyber Defense."

1.4 Objective

The objective of G0rKing is to automate the Google dorking process and offers a way to save the results in a text file for easy reference and reporting purposes. It can also automate the process of checking one URL against multiple dorks and can hence provide a quick health check.

How does Google dorking work?

Google uses the process called crawling to index new or latest pages. The program that does the crawling is called Googlebot. Googlebot uses an algorithmic process wherein computer programs determine which sites to crawl, and how many pages to fetch from each site. In Google dorking, the search engines crawl various data links, index page titles, web page contents, and store the information which will serve the purpose of search queries efficiently. Here the crawlers will bring various information out to the public, although the content owners did not intend to reveal it. The main aim of Google dorking is to locate useful information using techniques that are already provided by the search engine but in different ways.

1.4.1 Types of crawling

There are two types of crawling used by Googlebots. They are as follows –

Deep crawl: When Googlebots fetches a page, it selects all the links appearing on the page and adds them to a queue for further crawling. By collecting all the links, Googlebots covers a wide reach of the web.

Fresh crawl: Google keeps scanning and rescanning the popular and frequently changing web pages regularly. This is done at a rate highly proportional to how often the content of the pages changes.

Advantages: There are various advantages of Google dorking. Google dorking is effective as it indexes vast amounts of information in numerous formats, and that collection of data or information is growing every minute. Google can also index images, videos, all sorts of file types such as PDF, PPT, etc. All the information provided by Google is stored in large numbers, and we only need to know how to search for that data.

1.5 Technical Details

Google dorking resources: There are numerous resources about the process of Google dorking, but one of the best resources is about Google Hacking Database (GHDB). The GHDB is a collection of Google hacking search terms that allows the search of sensitive data hosted on any server or web application.

The GHDB was launched by Johnny Long in the year 2000, with an aim to serve penetration testers. In 2010, GHDB was adopted as a part of exploit-db.com. At that time, its scope was also increased beyond the Google search engine and included other search engines including Microsoft Bing. In simple terms, any user can refer to the GHDB to create search engine queries, which could find insecure web resources, such as configuration files or databases, which have been intentionally or unintentionally indexed by the search engine.

1.5.1 Google dorking techniques

Some of the Google dorking techniques are as follows:

Site mapping: To find every web page Google has crawled for a specific site, use the site: operator. Consider the following query:

```
site:http://www.microsoft.com
```

This query searches for the word Microsoft, restricting the search to the http://www.microsoft.com website. How many pages on the Microsoft web server contain the word Microsoft? According to Google, all of them!. Google searches not only the content of a page but the title and URL as well. The word Microsoft appears in the URL of every page on http://www.microsoft.com. With a single query, an attacker gains a rundown of every web page on a site cached by Google.

Finding directory listings: Directory listings provide a list of files and directories in a browser window instead of the typical text and graphics mix generally associated with web pages. These pages offer a great environment for deep information gathering. Several alternate queries provide more accurate results:

```
1  intitle:index.of ``parent directory''
2  intitle:index.of name size
```

These queries indeed provide directory listings by not only focusing on the index in the title, but on keywords often found inside directory listings, such as parent directory, name, and size. This search can be combined with other searches to find files or directories located in directory listings.

Versioning: Obtaining the web server software/version. The exact version of the web server software running on a server is one piece of information an attacker needs before launching a successful attack against that web server. If an attacker connects directly to that web server, the HTTP (web) headers from that server can provide this essential information. It's possible, however, to retrieve similar information from Google's cache without ever connecting to the target server under investigation. One method involves using the information provided in a directory listing. For example:

```
antitle:index.of server.at
```

This query focuses on the term index of in the title and server at appearing at the bottom of the directory listing. This type of query can also be pointed at a particular web server.

1.6 Project SaUR0N – One Tool to Search Them All

Project SaUR0N, now officially adopted by OWASP foundation as Project G0rking [62], is an open-source tool [10] to automate the Google dork search process. It can be used to search for any already indexed and publicly available information across the Internet. Pentesters and security researchers can use this tool to find security holes in the code of a website or associated resources accessible online. It uses the advanced operators available with the Google search engine. Figure 1.1 shows the Linux terminal for Project SaUR0N.

1.6.1 Project deliverables

- Project website [56] for hosting the educational content.
- Google, PyFiglet, and TQDM for basic requirements of the tool [10].
- Google dorking usage and easy reference in a text file.
- G0rKing aka SaUR0N is a Google dorking tool written in Python3.

Figure 1.1   Linux terminal showing the starting banner for Project SaUR0N.

Figure 1.2　Home page of the website sauron.in.

One of the aims of the project was to promote awareness about the overall concept and importance of reconnaissance and automate the reconnaissance process by using the Google dork search process. The project website [56] was created to spread this awareness by educating the users about Google dorking. The "Home Page" of the websites introduces the users to the concept of Google dorking through an easy-to-understand video. Then it describes how Google dorking works. It then covers content about various types of crawling used by Google, namely Deep crawl and Fresh crawl. The homepage finally tells about the advantages of Google dorking. Figure 1.2 shows the webpage of website sauron.in.

The "Technical Details" section of the website talks about the Google Hacking Database (GHDB), and various techniques for Google dorking, which includes site mapping, finding directory listings, and versioning. The "Project SaURON" page describes all the key attributes of the project and covers the process of running the project. Finally, "Meet the Team" page provides a brief introduction about the core team members involved in the project.

1.7 Project Requirements Packages

For its operations, the tool has some dependencies on three custom-built packages. These packages (Google, PyFiglet, and TQDM) are available on the same GitHub repository. A brief description is provided in this section.

Google: Python bindings to the Google search engine. Developers can use the Python package "google" to get results of Google searches from within the Python scripts. In case of multiple results, the package provides the entire list. The official link for the independent package: https://pypi.org/project/google/#description.

PyFiglet: Pure-Python FIGlet implementation. The FIGLet program is used to create large characters out of ordinary screen characters. It accepts ASCII text as input and renders it in ASCII art fonts. The official link for the independent package: https://pypi.org/project/pyfiglet/.

TQDM: Fast, extensible progress meter. This project allows developers to display a smart progress meter for any command in execution. Searching Google for several specific keywords and metadata may often be time taking, and the progress bar provides an indication of the expected time duration. The official link for the independent package: https://pypi.org/project/tqdm/.

1.8 DorkingGuide – Tool User Manual

The project deliverables also include a usage manual, that provides the list and explanation of Google dorking queries. Like most search engines, Google's search engine is also programmed to accept several advanced "filters" or "prefix operators" to fine-tune its searchers. This user manual provides the list of these filters and prefix operators that can be used to perform advanced searches across the web.

1.9 The Tool – G0rKing aka SaUR0N

G0rking is an open-source tool written in Python3. This makes it compatible with most of the Linux flavors available in the market. This is a command-prompt-based tool to help automate Google dorking. This is available on the GitHub repository, and available to everyone for download for free. Users can use this tool for two purposes: for performing Google dork searches, and for URL probing. Details about these are provided in later sections.

Installation prerequisites: Installation of this tool requires a few package files, which are listed in the document titled "requirements.txt" available in the same GitHub repository.

Operation: Running this tool requires no specific skill. Once installed, the tool provides on-screen instructions about the expected inputs in simple English language.

How to run? The entire process of installation and execution can be performed using the below commands on your Linux terminal.

```
1 $ git clone https://github.com/BlueVirtualNerds/SaUR0N
2 $ cd SaUR0N
3 $ pip3 install -r requirements.txt
4 $ python3 sauron
```

The above commands will download the project from its Github repository, install the packages required for the project, and then finally run the project. The project can be executed for the following two purposes:

- **Simple Google dork searches:** In this, users provide some search criteria, like a keyword or some metadata, and the project scans the web to find out all the web resources meeting the specified search criteria.
- **URL probing:** In this, the user can provide a specific URL, and the tool performs a scan on that URL to identify any known vulnerabilities or exposed resources related to that URL.

1.9.1 For simple google dorking (search)

When executed, the project provides three options to choose. For a simple Google dork search, users need to select option 1 in the starting banner screen. Figure 1.3 shows the corresponding banner.

While performing a web scan, the tool provides an option to save all the results into a file for a later probe. Saving the results has been kept optional, as the results may often include sensitive resources and links, which may lead to

Figure 1.3 Project SaUR0N starting banner (select option 1 for Google dork search).

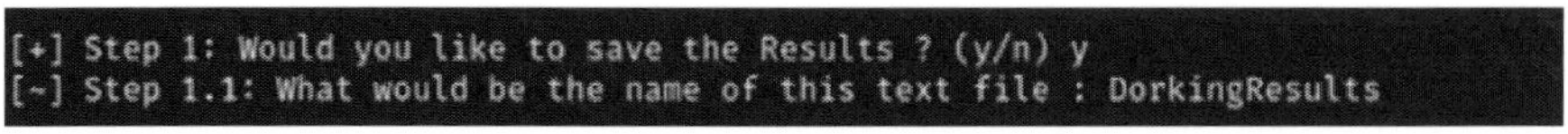

```
[+] Step 1: Would you like to save the Results ? (y/n) y
[~] Step 1.1: What would be the name of this text file : DorkingResults
```

Figure 1.4   Confirmation message about saving the results as a text file.

```
[+] Step 2: Enter Your Dork Search Query/String: filetype:pdf "Security Report"
```

Figure 1.5   Using the dork search query/string.

actual exploitation of the data. If the user wishes to save the search results, they are asked to provide a name for the file. Figure 1.4 shows the corresponding confirmation banner.

Users are requested to provide the search string or query, and other relevant parameters. For example, a user may wish to scan all the web resources, and search for all the PDFs (filetype:pdf), which have the word "Security Report." More possible options are provided in the "Dorking Queries" section below. Figure 1.5 shows the corresponding confirmation banner about the number of search results to be saved.

The tool then asks users to enter the number of search results to be scanned/displayed. In the case where a high number of search results are expected, restricting the number of results could help save time and resources out of lengthy processing.

As soon as this is done, the tool starts scanning for the matching patterns specified in the search query and then displays or saves the results, based on the chosen options.

1.9.2 For URL probing

For URL probing, user's need to select option 2 in the starting banner screen. Figures 1.6–1.8 show the corresponding banners for option 2. Similar to the process during simple Google dorking for search, the tool provides an option to save all the results into a file for a later probe. If the user wishes to save the results, they can say "y" and then provide the filename to be saved. The tool then asks for the URL that the user wishes to probe.

```
[+] Step 3: How many search results do you want to display/save ?: 10
```

Figure 1.6   Confirmation about the number of search results to save.

Figure 1.7 Project SaURON starting banner (Select option 2 or URL Probing).

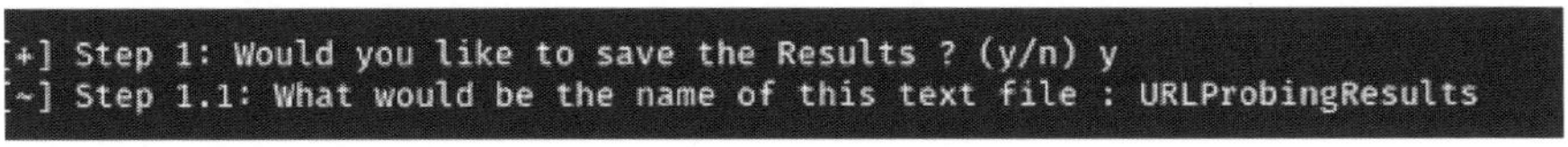

Figure 1.8 To save the results, say "y" and then provide the name for the file. To proceed without saving the results, select "n".

Figure 1.9 Enter the URL domain to be probed (without www). For example, you can enter sauron.in.

As soon as this is done, the tool starts scanning for the resources related to the provided domain name and then displays or saves the results, based on the chosen options.

1.10 Dorking Queries

When framing the search queries, users can choose among the long list of keywords and flags.

1.10.1 Guide

Cache: If you include other words in the query, Google will highlight those words within the cached document. For instance, [cache:www.google.com

web] will show the cached content with the word "web" highlighted. This functionality is also accessible by clicking on the "Cached" link on Google's main results page. The query **[cache:]** will show the version of the web page that Google has in its cache. For instance, [cache:www.google.com] will show Google's cache of the Google homepage. Note there can be no space between the "cache:" and the web page URL. The following listing shows the corresponding results.

```
1  link: The query [link:] will list webpages that have links
      to the specified webpage. For instance, [link:www.google
      .com] will list webpages that have links pointing to the
      Google homepage. Note there can be no space between the
      "link:" and the web page URL.
2  -----------------------------------------------------------
3  related: The query [related:] will list web pages that are "
      similar" to a specified web page. For instance, [related
      :www.google.com] will list web pages that are similar to
      the Google homepage. Note there can be no space between
      the "related:" and the web page URL.
4  -----------------------------------------------------------
5  info: The query [info:] will present some information that
      Google has about that web page. For instance, [info:www.
      google.com] will show information about the Google
      homepage. Note there can be no space between the "info:"
      and the web page URL.
6  -----------------------------------------------------------
7  define: The query [define:] will provide a definition of the
      words you enter after it, gathered from various online
      sources. The definition will be for the entire phrase
      entered (i.e., it will include all the words in the
      exact order you typed them).
8  -----------------------------------------------------------
9  stocks: If you begin a query with the [stocks:] operator,
      Google will treat the rest of the query terms as stock
      ticker symbols, and will link to a page showing stock
      information for those symbols. For instance, [stocks:
      intc yhoo] will show information about Intel and Yahoo.
      (Note you must type the ticker symbols, not the company
      name.)
10 -----------------------------------------------------------
```

11 site: If you include [site:] in your query, Google will
   restrict the results to those websites in the given
   domain. For instance, [help site:www.google.com] will
   find pages about help within www.google.com. [help site:
   com] will find pages about help within .com URLs. Note
   there can be no space between the "site:" and the domain
   .

12 --

13 allintitle: If you start a query with [allintitle:], Google
   will restrict the results to those with all of the query
   words in the title. For instance, [allintitle: google
   search] will return only documents that have both "
   google" and "search" in the title.

14 --

15 intitle: If you include [intitle:] in your query, Google
   will restrict the results to documents containing that
   word in the title. For instance, [intitle:google search]
   will return documents that mention the word "google" in
   their title, and mention the word "search" anywhere in
   the document (title or no). Note there can be no space
   between the "intitle:" and the following word. Putting [
   intitle:] in front of every word in your query is
   equivalent to putting [allintitle:] at the front of your

16 --

17 query: [intitle:google intitle:search] is the same as [
   allintitle: google search].

18 --

19 allinurl: If you start a query with [allinurl:], Google will
   restrict the results to those with all of the query
   words in the URL. For instance, [allinurl: google search
   ] will return only documents that have both "google" and
   "search" in the URL. Note that [allinurl:] works on
   words, not URL components. In particular, it ignores
   punctuation. Thus, [allinurl: foo/bar] will restrict the
   results to pages with the words "foo" and "bar" in the
   URL, but won't require that they be separated by a slash
   within that URL, that they be adjacent, or that they be
   in that particular word order. There is currently no
   way to enforce these constraints.

20 --

21 inurl: If you include [inurl:] in your query, Google will
   restrict the results to documents containing that word
   in the URL. For instance, [inurl:google search] will
   return documents that mention the word "google" in their
   URL, and mention the word "search" anywhere in the
   document (url or no). Note there can be no space between
   the "inurl:" and the following word. Putting "inurl:"
   in front of every word in your query is equivalent to
   putting "allinurl:" at the front of your query: [inurl:
   google inurl:search] is the same as [allinurl: google
   search].

22 --

23 Nina Simone intitle:"index.of" "parent directory" "size" "
   last modified" "description" I Put A Spell On You (mp4|
   mp3|avi|flac|aac|ape|ogg) -inurl:(jsp|php|html|aspx|htm|
   cf|shtml|lyrics-realm|mp3-collection) -site:.info

24 Bill Gates intitle:"index.of" "parent directory" "size" "
   last modified" "description" Microsoft (pdf|txt|epub|doc
   |docx) -inurl:(jsp|php|html|aspx|htm|cf|shtml|ebooks|
   ebook) -site:.info

25 parent directory /appz/ -xxx -html -htm -php -shtml -
   opendivx -md5 -md5sums

26 parent directory DVDRip -xxx -html -htm -php -shtml -
   opendivx -md5 -md5sums

27 parent directory Xvid -xxx -html -htm -php -shtml -opendivx
   -md5 -md5sums

28 parent directory Gamez -xxx -html -htm -php -shtml -opendivx
   -md5 -md5sums

29 parent directory MP3 -xxx -html -htm -php -shtml -opendivx -
   md5 -md5sums

30 parent directory Name of Singer or album -xxx -html -htm -
   php -shtml -opendivx -md5 -md5sums

31 filetype:config inurl:web.config inurl:ftp

32 "Windows XP Professional" 94FBR

33 ext:(doc | pdf | xls | txt | ps | rtf | odt | sxw | psw |
   ppt | pps | xml) (intext:confidential salary | intext:"
   budget approved") inurl:confidential

34 ext:(doc | pdf | xls | txt | ps | rtf | odt | sxw | psw |
   ppt | pps | xml) (intext:confidential salary | intext:"
   budget approved") inurl:confidential

35 ext:inc "pwd=" "UID="

36 ext:ini intext:env.ini

37 ext:ini Version=... password

38 ext:ini Version=4.0.0.4 password

39 ext:ini eudora.ini

40 ext:ini intext:env.ini

41 ext:log "Software: Microsoft Internet Information Services *.*"

42 ext:log "Software: Microsoft Internet Information

43 ext:log "Software: Microsoft Internet Information Services *.*"

44 ext:log \"Software: Microsoft Internet Information Services *.*\"

45 ext:mdb   inurl:*.mdb inurl:fpdb shop.mdb

46 ext:mdb inurl:*.mdb inurl:fpdb shop.mdb

47 ext:mdb inurl:*.mdb inurl:fpdb shop.mdb

48 filetype:SWF SWF

49 filetype:TXT TXT

50 filetype:XLS XLS

51 filetype:asp   DBQ=" * Server.MapPath("*.mdb")

52 filetype:asp "Custom Error Message" Category Source

53 filetype:asp + "|ODBC SQL"

54 filetype:asp DBQ=" * Server.MapPath("*.mdb")

55 filetype:asp DBQ=\" * Server.MapPath(\"*.mdb\")

56 filetype:asp "Custom Error Message" Category Source

57 filetype:bak createobject sa

58 filetype:bak inurl:"htaccess|passwd|shadow|htusers"

59 filetype:bak inurl:\"htaccess|passwd|shadow|htusers\"

60 filetype:conf inurl:firewall -intitle:cvs

61 filetype:conf inurl:proftpd. PROFTP FTP server configuration file reveals

62 filetype:dat "password.dat

63 filetype:dat \"password.dat\"

64 filetype:eml eml +intext:"Subject" +intext:"From" +intext:"To"

65 filetype:eml eml +intext:\"Subject\" +intext:\"From\" + intext:\"To\"

66 filetype:eml eml +intext:"Subject" +intext:"From" +intext:"To"

67 filetype:inc dbconn

68 filetype:inc intext:mysql_connect

69 filetype:inc mysql_connect OR mysql_pconnect

70 filetype:log inurl:"password.log"

71 filetype:log username putty PUTTY SSH client logs can reveal usernames

72 filetype:log "PHP Parse error" | "PHP Warning" | "PHP Error"

73 filetype:mdb inurl:users.mdb

74 filetype:ora ora

75 filetype:ora tnsnames

76 filetype:pass pass intext:userid

77 filetype:pdf "Assessment Report" nessus

78 filetype:pem intext:private

79 filetype:properties inurl:db intext:password
80 filetype:pst inurl:"outlook.pst"
81 filetype:pst pst -from -to -date
82 filetype:reg reg +intext:"defaultusername" +intext:"
      defaultpassword"
83 filetype:reg reg +intext:\"defaultusername\" +intext:\"
      defaultpassword\"
84 filetype:reg reg +intext:Ãć? WINVNC3Ãć?
85 filetype:reg reg +intext:"defaultusername" +intext:"
      defaultpassword"
86 filetype:reg reg HKEY_ Windows Registry exports can reveal
87 filetype:reg reg HKEY_CURRENT_USER SSHHOSTKEYS
88 filetype:sql "insert into" (pass|passwd|password)
89 filetype:sql ("values * MD5" | "values * password" | "values
      * encrypt")
90 filetype:sql (\"passwd values\" | \"password values\" | \"
      pass values\")
91 filetype:sql (\"values * MD\" | \"values * password\" | \"
      values * encrypt\")
92 filetype:sql +"IDENTIFIED BY" -cvs
93 filetype:sql password
94 filetype:sql password
95 filetype:sql "insert into" (pass|passwd|password)
96 filetype:url +inurl:"ftp://" +inurl:";@"
97 filetype:url +inurl:\"ftp://\" +inurl:\";@\"
98 filetype:url +inurl:"ftp://" +inurl:";@"
99 filetype:xls inurl:"email.xls"
100 filetype:xls username password email
101 index of: intext:Gallery in Configuration mode
102 index.of passlist
103 index.of perform.ini mIRC IRC ini file can list IRC
      usernames and
104 index.of.dcim
105 index.of.password
106 intext:" -FrontPage-" ext:pwd inurl:(service | authors |
      administrators | users)
107 intext:""BiTBOARD v2.0" BiTSHiFTERS Bulletin Board"
108 intext:"# -FrontPage-" ext:pwd inurl:(service | authors |
      administrators | users) "# -FrontPage-" inurl:service.
      pwd
109 intext:"#mysql dump" filetype:sql
110 intext:"#mysql dump" filetype:sql 21232
      f297a57a5a743894a0e4a801fc3
111 intext:"A syntax error has occurred" filetype:ihtml
112 intext:"ASP.NET_SessionId" "data source="

113 intext:"About Mac OS Personal Web Sharing"
114 intext:"An illegal character has been found in the statement " -"previous message"
115 intext:"AutoCreate=TRUE password=*"
116 intext:"Can't connect to local" intitle:warning
117 intext:"Certificate Practice Statement" filetype:PDF | DOC
118 intext:"Certificate Practice Statement" inurl:(PDF | DOC)
119 intext:"Copyright (c) Tektronix, Inc." "printer status"
120 intext:"Copyright Âl' Tektronix, Inc." "printer status"
121 intext:"Emergisoft web applications are a part of our"
122 intext:"Error Diagnostic Information" intitle:"Error Occurred While"
123 intext:"Error Message : Error loading required libraries."
124 intext:"Establishing a secure Integrated Lights Out session with" OR intitle:"Data Frame - Browser not HTTP 1.1 compatible" OR intitle:"HP Integrated Lights-
125 intext:"Fatal error: Call to undefined function" -reply -the -next
126 intext:"Fill out the form below completely to change your password and user name. If new username is left blank, your old one will be assumed." -edu
127 intext:"Generated   by phpSystem"
128 intext:"Generated by phpSystem"
129 intext:"Host Vulnerability Summary Report"
130 intext:"HostingAccelerator" intitle:"login" +"Username" -"news" -demo
131 intext:"IMail Server Web Messaging" intitle:login
132 intext:"Incorrect syntax near"
133 intext:"Index of" /"chat/logs"
134 intext:"Index of /network" "last modified"
135 intext:"Index of /" +.htaccess
136 intext:"Index of /" +passwd
137 intext:"Index of /" +password.txt
138 intext:"Index of /admin"
139 intext:"Index of /backup"
140 intext:"Index of /mail"
141 intext:"Index of /password"
142 intext:"Microsoft (R) Windows * (TM) Version * DrWtsn32 Copyright (C)" ext:log
143 intext:"Microsoft CRM : Unsupported Browser Version"
144 intext:"Microsoft Âl' Windows * âDĆ Version * DrWtsn32 Copyright Âl'" ext:log
145 intext:"Network Host Assessment Report" "Internet Scanner"
146 intext:"Network Vulnerability   Assessment Report"
147 intext:"Network Vulnerability Assessment Report"

```
148 intext:"Network Vulnerability Assessment Report"
    æIJňæŨĞæÌĕèĞł pc007.com
149 intext:"SQL Server Driver][SQL Server|Line 1: Incorrect
    syntax near"
150 intext:"Thank you for your order"  +receipt
151 intext:"Thank you for your order" +receipt
152 intext:"Thank you for your purchase" +download
153 intext:"The following report contains confidential
    information" vulnerability -search
154 intext:"phpMyAdmin MySQL-Dump" "INSERT INTO" -"the"
155 intext:"phpMyAdmin MySQL-Dump" filetype:txt
156 intext:"phpMyAdmin" "running on" inurl:"main.php"
157 intextpassword | passcode)  intextusername | userid | user)
    filetype:csv
158 intextpassword | passcode) intextusername | userid | user)
    filetype:csv
159 intitle:"index of" +myd size
160 intitle:"index of" etc/shadow
161 intitle:"index of" htpasswd
162 intitle:"index of" intext:connect.inc
163 intitle:"index of" intext:globals.inc
164 intitle:"index of" master.passwd
165 intitle:"index of" master.passwd 007çȚţèĎŚèţĎèőŕ
166 intitle:"index of" members OR accounts
167 intitle:"index of" mysql.conf OR mysql_config
168 intitle:"index of" passwd
169 intitle:"index of" people.lst
170 intitle:"index of" pwd.db
171 intitle:"index of" spwd
172 intitle:"index of" user_carts OR user_cart
173 intitle:"index.of *" admin news.asp configview.asp
174 intitle:("TrackerCam Live Video")|("TrackerCam Application
    Login")|("Trackercam Remote") -trackercam.com
175 intitle:("TrackerCam Live Video")|("TrackerCam Application
    Login")|("Trackercam Remote") -trackercam.com
176 inurl:admin inurl:userlist Generic userlist files

1 {
2 inurl:php?=id1
3 inurl:index.php?id=
4 inurl:trainers.php?id=
5 inurl:buy.php?category=
6 inurl:article.php?ID=
7 inurl:play_old.php?id=
8 inurl:declaration_more.php?decl_id=
9 inurl:pageid=
```

```
10  inurl:games.php?id=
11  inurl:page.php?file=
12  inurl:newsDetail.php?id=
13  inurl:gallery.php?id=
14  inurl:article.php?id=
15  inurl:show.php?id=
16  inurl:staff_id=
17  inurl:newsitem.php?num= andinurl:index.php?id=
18  inurl:trainers.php?id=
19  inurl:buy.php?category=
20  inurl:article.php?ID=
21  inurl:play_old.php?id=
22  inurl:declaration_more.php?decl_id=
23  inurl:pageid=
24  inurl:games.php?id=
25  inurl:page.php?file=
26  inurl:newsDetail.php?id=
27  inurl:gallery.php?id=
28  inurl:article.php?id=
29  inurl:show.php?id=
30  inurl:staff_id=
31  inurl:newsitem.php?num=
32  }
```

Listing 1.1 Use of the popular search string to find vulnerable websites.

For additional details and examples, one can refer to ''https://github.com /BlueVirtualNerds/SaUR0N/blob/main/DorkingGuide''

1.11 Best Practices and Learnings

Following all possible security best practices had been the main strategy followed during the development of project SaUR0N. Since its inception, all the processes and deliverables of this project are provided due to security considerations. For managing all the project-related documentation, a dedicated centralized repository was created on Google Drive, having restricted access only for the core members of the team and the teachers and mentors. All the team meetings were properly documented right from the beginning.

In the same manner, all the deliverables of this project are provided with proper security considerations. In the same manner, all the deliverables of this project are provided with proper security considerations.

1.12 Website Security and Best Practices

All the web-facing assets like a website stand directly in the front line of cyberattacks and therefore need extra security considerations. During development, it is easy to miss out on certain aspects, which may result in chaos later. Therefore, several organizations and professionals recommend using some mature web-development framework or platform, instead of developing everything from scratch. Well-established cybersecurity frameworks are backed by a strategic approach. They are an outcome of detailed research on security aspects and are often developed while considering cyber incident response plans and application security checklists in mind. For the development of enterprise-size applications or a heavy-use e-commerce portal, a strategic approach is needed, but for small-sized websites with limited capabilities, a reliable web framework sufficiently provides all the required security norms.

Therefore, when developing the website [56] for Project SaUR0N, we used the Wix IaaS, a reliable vendor that takes care of security aspects and follows the security norms. This way, we were able to enjoy a secure web-hosting space, where infrastructure-related security aspects were handled by experts in that domain.

The website uses SSL certification to provide an extra layer of security to its content. It also implements the latest version of TLS (TLS 1.3), which provides several enhancements over its previous version (TLS 1.2) in terms of performance and security. For additional security, regular backups were taken of the website data. Additionally, third-party vulnerability assessments were performed, which helped in identifying any technical security loopholes like the use of old versions or components. Due to the use of a third-party framework for development, there were some restrictions on the customization of cookies, which could have resulted in some risks. However, the entire website was designed to work with HTTP read-related operations (GET requests) only. By not using any write-related operations (POST requests) to the server, several vulnerabilities related to cookies were taken care of.

1.13 Tool: SaUR0N

Project SaUR0N tool is written in Python, which is considered one of the safest languages. The high severity vulnerabilities in the past 5 years are 15% on average, the lowest among the other languages. Additionally, we used proper tricks and development guidelines to ensure security. The tool has been kept open-source for everyone to review and find and fix bugs. The tools provide an option to the users if they wish to save sensitive data about vulnerabilities or not. This allows users to avoid any risks in case they wish not to save their search results. The tool is fully customizable. Users are provided with an option to add more queries as per the requirements for the URL probe part.

1.14 GitHub Repository

Project SaUR0N is an open-source work, and its entire code and libraries are available on GitHub for anyone to review (https://github.com/BlueVir tualNerds/SaUR0N). While working with GitHub, all standard guidelines were followed. During the development phase, the project repository was secured via 2FA. The credentials were shared using one-time read messaging services, in which messages carrying sensitive information (in our case credentials) were destroyed after one-time use. For any users visiting the GitHub project directly (and not via the project's education website) the Readme.md provides all the necessary details. This project is also having an active security policy (SECURITY.md), inviting contributions from all users, and provides warnings about any possible misuse of the tool and associated risks and accountabilities.

The GitHub repository is also using CodeQL – the code analysis engine developed by GitHub. This helps in automating the security checks of the hosted code. In case there is any vulnerability or risk identified, the build will automatically fail, thus securing the system from any supply chain-kind attacks.

Future prospects: In the future, it can include a scoring mechanism (like VirusTotal) to quantify the risk. The same functionalities can be delivered via an API-based and a GUI-based offering in the future. Moreover, after adoption by the OWASP foundation, there are enormous ways in which this project can be taken further.

2

OSS Known Vulnerability Scanner – Helping Software Developers Detect Third-Party Dependency Vulnerabilities in Real Time

Om Mishra and Ria Sarkar

Independent Researcher, India
E-mail: mishra.om@live.com; riasarkar1999@gmail.com

Abstract

A major problem every developer is facing today is the unavoidable dependencies on external pieces of codes. Current state of software industries shows that 98% of all applications rely on some form of open-source dependency and 84% of these contain vulnerabilities. Additionally, the average time taken for open-source vulnerability detection is around 22 years. The fact that OWASP considers this among their top 10 vulnerabilities ever since 2013 represents the severity of this issue. CVE databases are maintained, which contain in them a list of publicly known cybersecurity vulnerabilities. Currently, the number of open-source dependency checkers are only compatible with a few languages. On the other hand, commercial tools may be too expensive for start-up companies to use. To address this problem, our aim with this application is to develop a solution that will scan all open-source libraries used in software development, check the national vulnerability database's CVE database to find any known vulnerabilities and alert the developer in real time. We have developed the core features that can solve this problem in a very simple way by developing a standalone version with three components: a scanner that runs on the developer's machine and collects the evidence of use of any open-source libraries, an analyzer which processes the output generated by the scanner and checks for any vulnerabilities by calling NVD's CVE API and a dashboard to show the results generated by the analyzer. The software is currently compatible with projects made using Maven and Gradle and is hosted locally on the user's system. The error rates were found to be low, with a false positive of 0% and a false negative of 5%.

Keywords: Open-source vulnerability scanner, open-source software (OSS), OWASP, National Vulnerability database (NVD), Common Vulnerabilities and Exposures (CVE), Common Vulnerability Scoring System (CVSS).

2.1 Introduction

Dependencies are external pieces of codes used in a software to implement a given functionality. More often than not, software development involves reusing these existing pieces of codes. There are significant advantages of using dependencies during software development. These range from the gross reduction of time and resources needed such as elimination of the need to write and test every functionality from scratch to being able to create complex codes for highly specific applications easily by incorporating existing pieces of software codes. It also reduces code redundancy and facilitates sharing of software codes between developers through platforms such as GitHub. Software dependencies can be of two types. Direct dependencies are those that are directly called upon by the software to directly enable a particular functionality of the software. On the other hand, indirect or translative dependencies are those that are called upon by direct dependencies of the software instead of the software itself. While a developer may be aware of the direct dependencies associated with their software codes, the translative dependencies often go unnoticed.

The security risks associated with utilizing dependencies in a software code makes these vulnerabilities among the most exploited in the information and technology sector. Dependency on external pieces of codes can lead to accidental flaws such as the use of untested codes, implementing them in a manner different than that intended by the developer, or using older versions of a code. However, more dangerous are intentional vulnerabilities that can get incorporated through the use of compromised or malicious codes such as introducing backdoors leading to unauthorized access. As the name suggests, it also causes a dependency of the software on external code, meaning that the independence of the software to function as intended by the developer is greatly reduced and always relies on the status of these external and often uncontrollable sources.

The open-source vulnerability scanner scans the source code to collect the evidence of using any OSS library.Once the concrete evidence is established,

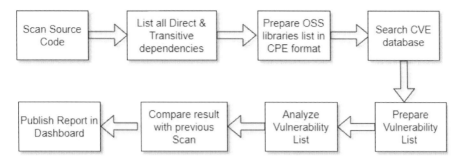

Figure 2.1 Workflow of developed open-source vulnerability scanner

it will further scan for all transitive dependencies and prepare a list.Once the OSS library list is prepared the tool will start checking NVD for any known vulnerability. It fetches all details of the vulnerability from different sources to prepare the dashboard. Figure 2.1 shows the information flow.

2.2 Background

Information technology has exploded in the past two decades. However, with such advancements, significant security threats are following. In today's world, writing every aspect of a code from scratch is not only resource-intensive and counterintuitive but also impractical and, in many cases, impossible if one wants to create an industry standard product. The advantages and widespread use associated with external dependencies alongside the threats they pose make these a necessary evil in the process of software development. It is therefore essential to identify and mitigate the risks associated with the use of external dependencies through the use of proactive preventive measures such as dependency management procedures. To combat this problem, Common Vulnerabilities and Exposures (CVE) databases are maintained, which contain in them a list of publicly known cybersecurity vulnerabilities [59].

The CVE database is maintained by the MITRE organization and is aimed to identify, define, and catalog publicly disclosed cybersecurity vulnerabilities. There is one CVE Record for each vulnerability in the catalog. The vulnerabilities are discovered then assigned and published [58]. The scoring system of the CVE database generates a score using the Common Vulnerability Scoring System (CVSS). CVSS consists of three metric groups: base, temporal, and environmental. The base group represents the intrinsic qualities of a vulnerability that are constant over time and across user environments, the temporal group reflects the characteristics of a vulnerability that change over time, and the environmental group represents the characteristics of a vulnerability that are unique to a user's environment. The base metrics produce a score ranging from 0 to 10, which can then be modified by scoring the temporal and environmental metrics [32]. Tables 2.1 and 2.2 explain the three metric groups, their associated components, and metric value of each component.

2.3 Problem Statement

The current state of software industries shows that 98% of all applications rely on some form of open-source dependency and 84% of these contain vulnerabilities. Additionally, the average time taken for open-source vulnerability detection is around 22 years [2]. The severity of this issue is represented by the fact that the open web application security project or OWASP, the leading non-profit organization formed to improve security of software, considers this among their top 10 vulnerabilities ever since 2013, wherein vulnerable and outdated components is ranked as the sixth most critical web application security risk in 2021 [1]. To make matters worse, the number of open-source dependency checkers currently available is only compatible with a few languages [5].

2.4 Tool Architecture

OSS known vulnerability scanner contains three major components which are as follows:

Scanner: The scanner walks the files and directories specified in command line parameter and collects list of OSS dependencies direct and transitive.

Analyzer: The analyzer searches the CVE database and prepare vulnerable OSS library list. It also analyzes the vulnerability and fetch other details from external source to suggest possible fixes.

Report dashboard: The reporter generates a vulnerability report in the form of a live dashboard.

The application follows the SaaS architecture and only the scanner component resides on either developer system as IDE plugin or on CI (continuous integration) server. The scanner sends the list of OSS library for analysis to cloud hosted application. Figure 2.2 shows the architecture of OSS vulnerability scanner.

Table 2.1 Metric groups used by the CVSS to generate a vulnerability score.

Metric	Submetric	Value	Description
Base	Attack Vector (AV)	Network (N) Adjacent (A) Local (L) Physical (P)	Context of exploitation. The value will be larger the more logically and physically remote an attacker can be in order to exploit the vulnerable component.
	Attack Complexity (AC)	Low (L) High (H)	Conditions that must exist in order to exploit the vulnerability. The base score is greatest for the least complex attacks.
	Privileges Required (PR)	None (N) Low (L) High (H)	The level of privileges an attacker must possess before successfully exploiting the vulnerability. The base score is greatest if no privileges are required.
	User Interact (UI)	None (N) Required (R)	Requirement for a human user, other than the attacker. The base score is greatest when no user interaction is required.
	Scope (S)	Unchanged (U) Changed (C)	Whether a vulnerability in one vulnerable component impacts resources in components beyond its security scope, the subjects and objects under the jurisdiction of a single security authority. The base score is greatest when a scope change occurs.
	Confidentiality (C)	High (H) Low (L) None (N)	Impact to the confidentiality of the information resources managed by a software component due to a successfully exploited vulnerability. The base score is greatest when the loss to the impacted component is highest.
	Integrity (I)	High (H) Low (L) None (N)	Impact to integrity of a successfully exploited vulnerability. Integrity refers to the trustworthiness and veracity of information. The base score is greatest when the consequence to the impacted component is highest.
	Availability (A)	High (H) Low (L) None (N)	Impact to integrity of a successfully exploited vulnerability. Integrity refers to the trustworthiness and veracity of information. The base score is greatest when the consequence to the impacted component is highest.

2.5 Tool Implementation

This section covers how the various components are implemented and deployed for actual usage. The tool is implemented having three different components as mentioned in Section 2.4. Figure 2.3 shows the various technologies

Table 2.2   Metric groups used by the CVSS to generate a vulnerability score.

Metric	Submetric	Value	Description
Temporal	Exploit Code Maturity (E)	**Not Defined (X)** **High (H)** **Functional (F)** **Proof of Concept (P)** **Unproven (U)**	Likelihood of the vulnerability being attacked, and is typically based on the current state of exploit techniques, exploit code availability, or active, "in-the-wild" exploitation. The more easily a vulnerability can be exploited, the higher the vulnerability score.
	Remediation Level (RL)	**Not Defined (X)** **Unavailable (U)** **Workaround (W)** **Temporary (T)** **Official Fix (O)**	The availability of a remediation - an official patch or update against known vulnerabilities. The less official and permanent a fix, the higher the vulnerability score.
	Report Confindence (RC)	**Not Defined (X)** **Confirmed (C)** **Reasonable (R)** **Unknown (U)**	The degree of confidence in the existence of the vulnerability and the credibility of the known technical details. The more a vulnerability is validated by the vendor or other reputable sources, the higher the score.
Environmental	Security Requirements (CR, IR, AR)	**Not Defined (X)** **High (H)** **Medium (M)** **Low (L)**	These metrics enable the analyst to customize the CVSS score depending on the importance of the affected IT asset to a userâĂŹs organization, measured in terms of confidentiality, integrity, and availability.
	Modified Base Metrics	**Same as** **Base Metrics**	These metrics enable the analyst to override individual base metrics. Includes modified attack vector (MAV), modified attack complexity (MAC), modified privileges required (MPR), modified user interaction (MUI), modified scope (MS), modified confidentiality (MC), modified integrity (MI), Modified Availability (MA)

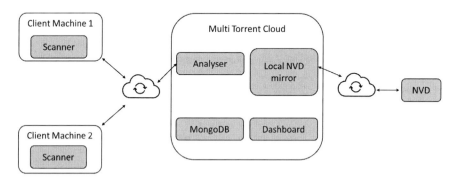

Figure 2.2   SaaS architecture of the OSS known vulnerability scanner

and tools used in our work. The component-wise implementations are as follow:

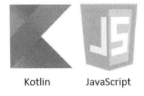

Kotlin   JavaScript

Vulnerability Databases

Maven   Gradle

CVE Search

MongoDB   Google Cloud Platform

Figure 2.3   Technologies and tools used to develop the tool.

2.5.1 Components and their implementations

Scanner: Scanner component is written in Kotlin [4] and its four variants make it suitable for various needs:

Command line utility: The simplest form of the scanner is a command line utility. To invoke the scanner, enter the following command in command shell: **java -jar scanner.jar [project path]**. This scanner can be added with various plugins as follows:

- **Maven plugin:** Just add the scanner as Maven plugin and it will automatically trigger when Maven build is invoked.
- **Gradle plugin:** Just add the scanner as Gradle plugin and it will automatically trigger when Gradle build is invoked.
- **Jenkin plugin:** Just add the scanner as Jenkin plugin and add the build job in Jenkin configuration file. Scanner will be invoked as per the configured stage in pipeline job.

Analyzer: The analyzer component can be hosted as SaaS application or on-premise.The analyzer is invoked as API call and takes the list of OSS libraries as input (in JSON format). The analyzer first look for CPE (Common Platform Enumeration) by grouping the vendor, product, and version of each OSS library and once the CPE is confirmed it looks for the CVE entry in NVD database. Each CVE entry contains:

- Description of the vulnerability or exposure,
- Common Vulnerability Scoring System (CVSS) score,
- List of the affected platforms identified by their Common Platform Enumeration (CPE).

The analyzer then prepares the report in JSON format as a result outcome.

Dashboard: A live reporting system was created using JavaScript, CSS, and HTML languages using the Visual Studio software [51]. The dashboard takes the output of the scanner in the form of the JSON object file and generates a live dashboard listing and comparing the total number of dependencies found. A vulnerability analyser depicts the total number of dependencies found and the relative proportion of high-risk ones through an intuitive pie chart. Additionally, a list of all the vulnerable dependencies found in the code is generated and displayed, containing all the important details that the developer might need such as the vulnerability score, risk category, description, links to the associated CVE, and CWE databases wherein the developer may find possible solutions such as an alternate dependency or an updated version of the code with higher security.

Figure 2.4 depicts the report generated after analysing a vulnerable software code downloaded from github. Out of the total 15 dependencies, nine had no known vulnerabilities, one posed a low-risk threat, two were medium and high-risk threats while one dependency was critically vulnerable.

2.6 Deployment

2.6.1 Enterprise deployment

The key challenge for enterprise deployment was how to assure the enterprise that no source code/IP is shared to our SaaS-hosted server. To ensure that we developed the scanner component as a plugin for IDE like eclipse, IntelliJ, etc. The scanner just looks for the OSS library and only send the list of libraries to the SaaS component. The SaaS component do check the vulnerability from NVD database. The NVD database is mirrored for faster analysis and its periodically sync every 2 hours.

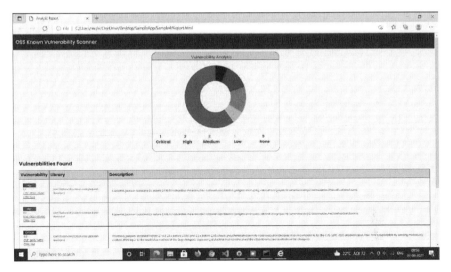

Figure 2.4   Live dashboard report generated following scanning of the software code.

Table 2.3   Validation result of tool.

No. of projects scanned	False positive	False negative
115	0	6

2.6.2 Standalone deployment

It is also possible to host entire application as on-premise offering if enterprise policy does not allow the cloud-based SaaS offering. Here are the components (scanner, analyzer, and dashboard) hosted in interned or in developer machine itself.

2.7 Tool Validation and Result

We validated the application by scanning 115 OSS projects downloaded from GitHub and analyzing the result manually. The result was very impressive and shown in Table 2.3.

2.8 Conclusion

The developed OSS known vulnerability scanner was found to be a reliable solution for software developers. It provides an alternative to expensive commercially available software for small-and medium-size enterprises as well as independent developers to help increase the security of their products by identifying and reporting exploitable vulnerabilities present in the software code based on the Common Vulnerabilities and Exposures database. The scanner currently supports software codes written using the Maven and Gradle platforms but can easily be modified to incorporate other programming languages. The low-false positive and false-negative results have proven the tool to be capable of generating reliable and valid results.

2.9 Acknowledgments

The authors are grateful to the Indian Institute of Technology, Kanpur, India and TalentSprint for providing this opportunity.

Part II

Malware Analysis

3

Detecting Malware using Machine Learning

**Partha Majumdar, Shyava Tripathi, Balaji Annamalai,
Senthil Jagadeesan, and Ranveer Khedar**

E-mail: partha.majumdar@hotmail.com; shyava.tripathi8@gmail.com;
Balajiannamalai@hotmail.com; jj_senthil75@yahoo.co.in;
ranveerkhedar@gmail.com

Abstract

The threat of malware looms large in today's world. It is of paramount importance that malware detection strategies evolve at an appropriate speed. The recent boost in exchange of images on social media as well as popularity of IoT devices are significant contributors of the current malware landscape. The present-day malware detection products are based on intensive manual effort and thus consume a lot of time to detect malware. It is a burning need to be able to detect malware in less time. This will not just save time and thus negate malicious intent of malware developers; it will also save a lot of money involved in malware detection.

Training machines to detect malware can significantly reduce the time required to detect malware. These techniques can be reasonably reliable because by following a consistent process, all but the outliers can be detected. Once machines can be made to learn the process of detecting malware, the time required to detect malware will reduce drastically. Also, as the manpower required in the process of malware detection using machines will be a fraction of what is required now, the net cost of detecting malware will reduce drastically.

In this chapter, we develop and evaluate two machine learning classifiers, the former capable of detecting malicious JPEG files with 99.9% accuracy and the latter capable of detecting and classifying malicious ELF files into malware categories with 87% accuracy.

Keywords: Malware, machine learning, JPEG, ELF, detection, classification, TF-IDF, Random Forest, confusion matrix, web service, Python.

3.1 Introduction

Malware threats continue to soar in numbers and diversify in functionality owing to the opportunities provided by technological advances. The attack surface is widening enormously everyday with the rise in data exchange and the highly personal interaction of technology with people's lives.

One very ubiquitous file exchanged by people across the globe are images. The advent of social media has boosted the exchange of images. JPEG is the most widely used image format, utilized by almost everyone, from individuals to schools and large enterprises. JPEG files, if exploited, could serve as a terrifying attack vector because of their extensive platform compatibility and expansive usage. JPEG files have many components that can be infected. This chapter deals with one such component, the EXIF (Exchangeable Image File Format) tags. EXIF tags are introduced into JPEG files in 2010 and are used to store meta data about the image stored in the JPEG file. A JPEG file with tampered EXIF tags is very difficult for users of the image file to detect. This is the incentive for malware developers to embed their malicious code in the EXIF tags. JPEG files can be cleaned of malware by converting the JPEG files to BMP files. BMP files do not store the EXIF tags present in the JPEG files.

Another increasingly exploitable attack surface in today's rapidly evolving malware landscape is that of Linux-based IoT (Internet of Things) devices and thus Linux binary files. The rise of connected objects with the evolution and revolution of IoT has led to a massive influx of Linux-based malware and attacks targeting IoT devices. The increasing number of Linux malware can be attributed to the infrequent patching, use of default passwords and almost negligible isolation of these devices which makes them extremely vulnerable to exploitation. Manual malware analysis is ineffective due to large number of such cases and so detection models based on machine learning approaches could serve valuable results. In this chapter, we propose two different machine learning solutions.

- In the first part, we present a machine learning solution aimed at efficiently detecting malicious JPEG files with an accuracy of 99.9%. The algorithms used are Decision Tree & Random Forest algorithms.
- In the second part, we introduce a machine learning solution based on the Random Forest algorithm, capable of detecting and classifying malicious ELF files into malware categories of Trojan, Backdoor, Virus, DDOS, and Botnet.

3.2 Detecting Malware in JPEG Files

The project tries to demonstrate that malware can be detected through supervised machine learning. Using machine learning for malware detection [39] has technical and commercial benefits. This is because malware detection in the current context is about constantly researching for new signatures developed by hackers and programming for detecting these signatures in files. This process is effort intensive and time intensive. The result is an expensive process for malware detection. If the machine could be made to detect malware, not only will the process of malware detection become less expensive due to less human effort involved, but the time required to detect new malware will reduce drastically.

3.2.1 JPEG file structure

JPEG files are compressed files used to store images. The JPEG files can be identified by a marker 0xFFD8 at the start of the file [33].

The information in the JPEG files can be classified into two types-Image Data and EXIF tags. EXIF tags are added to JPEG file format in 2010. The Image Data of JPEG files have the following segments:

1. Header
2. Quantization tables
3. Frame information

The Header segment contains the following data:

1. Identifier (OxJFIF for JPEG, etc.)
2. Version
3. Units
4. Density
5. Thumbnail

The frame information is a series of Huffman encoded tables containing the bit pattern of the image.

The important thing to note is that if any of above values is tampered with, the JPEG file will not render appropriately. So, the developers of malware do not tamper with this part of JPEG files to introduce malware.

3.2.2 EXIF tags

EXIF tags provide additional information to the JPEG files. The EXIF tags can be altered programmatically to alter the nature of the image stored in the JPEG files. For example, by altering the EXIF tag BrightnessValue, the brightness of the image can be altered. Similarly, by altering the EXIF tags ExifImageHeight and ExifImageWidth, the size of the image can be altered.

EXIF tags provide the facility for photograph editors to make enhancements/alterations to an image stored as a JPEG File. As these are values which can be altered, the developers of malware alter these tags to introduce their spurious code. One way to clean an infected JPEG file is to convert the JPEG file to a BMP file. BMP files can only contain the image information and does not support any EXIF tags. Converting to BMP files gets rid of the EXIF tags and thus the JPEG files get cleaned of malware. A JPEG file may contain EXIF tags or may not contain EXIF tags. JPEG files created before 2010 do not contain EXIF tags. If a JPEG file does not contain EXIF tags, then it can be safe to state that the JPEG file can be considered as a clean file.

So, for malware analysis, we need concentrating only on JPEG files which contain EXIF tags.

3.3 Dataset

The initial data for this project is obtained from C3i Lab of IIT, Kanpur. IITK provided 3124 clean JPEG files and 278 JPEG files with malware. As the total number of files are less, 2763 JPEG files are collected from our libraries of photographs. These files are transferred using Google Drive. Transferring using Google Drive is a round of check to ensure that these files are clean as Google Drive discards any JPEG files containing a malware. As the number of JPEG files containing malware is less, C3i Lab of IIT, Kanpur is approached. C3i Lab provided another 160 JPEG files containing malware. So, the total dataset contained 5887 clean JPEG files and 438 JPEG files with malware.

3.3.1 Dataset split – train and test set

To create the training set and test set, it is required that files be picked up at random and assigned to each set. To randomly pick up files, a shell script is written as shown in Listing 3.1.

```
1  linesInFile=$(wc -l < lst.txt)
2  echo $linesInFile
3  n_line=0
4  n_fileRequired=60
5  while read line
6  do
7    (( n_line == n_fileRequired )) && break
```

```
 8    rnd=$(( 1 + $RANDOM % $linesInFile ))
 9    sed -n "${rnd}p" lst.txt >> randomList.txt
10    ((n_line++))
11  done < lst.txt
```

Listing 3.1   FileMover.sh script.

The way this shell script is used is as follows:

1. All the files of a particular type are placed in a separate directory. For example, all the benign files are placed in a directory called **clean_jpeg**, all the malware files are placed in the directory **malicious_files**.
2. The list of files in each directory is created using the ls command and stored in a file called **lst.txt** (*ls > lst.txt*).
3. As we needed 10% of the files for the test set, the n_fileRequired variable in the script is set to the desired number. For example, 600 is about 10% of the total clean JPEG files. So, for the set of clean JPEG files, n_FileRequired could be set to 600.
4. Once the setting is made in the script as per Step 3, the shell script is run (**sh FileMover.sh**). (It is also possible that execute permission could have been given to the file **FileMover.sh** using the command **chmod u+x** and the file could have been run directly on the shell as **FileMover.sh**).
5. Running the command in Step 4 created the file **randomList.txt**. This file contained the names of the files randomly selected.
6. The file **randomList.txt** is updated in vi editor to create the move script for moving the files. The following commands are issued to generate the script.

```
1  a.   <Esc>:1,$s/^/mv"/
2  b.   <Esc>:1,$s/$/" \.\.\/Test\/Clean/
```

7. The updated file **randomList.txt** is run to move the files by issuing the command **sh randomList.txt**. This created a directory called Test or Clean for the clean JPEG files for the test set.
   Note: The directory **Test** had to be created at the appropriate location prior to executing Step 6.
8. These steps are repeated for all the types of files. So, now there are two sets of files of each type – one for training and one for testing.

Out of the total dataset, 5398 clean JPEG files and 415 malicious JPEG files are used for the training set and remaining, 292 clean JPEG files and 22 malicious JPEG files are used for the testing set.

3.4 Feature Extraction Strategies

The initial part of the project involves finding reliable libraries that can help in parsing the JPEG files. After parsing the JPEG files, we see two significant discoveries that not all the JPEG files in the obtained dataset are JPEG files and that many of the JPEG files did not contain EXIF tags [16]. Both these reasons reduced the dataset. This discovery also led to the following conclusions:

1. If the file is not a valid JPEG File, the software would just reject these files as not a part of the scope for this product.
2. If the file is a valid JPEG file but doese not contain EXIF tags, then the file will be classified as benign. (This is because in almost all cases it is not possible to transmit malware in a JPEG File without editing the EXIF tags).
3. Some of the JPEG files containing malware are found to be not JPEG files. It is possible as the file becomes damaged while it is being manipulated. The signature identifies JPEG files from the header information present in the JPEG files.
4. The other observation is that few of the malicious files in the dataset are found that do not contain any EXIF tags. A check is performed by uploading these files to Google Drive. The files got uploaded to Google Drive. However, when we tried to download these files from Google Drive, they were reported as errors.
5. Out of the JPEG files containing malware, one file is found that the Python program can not parse because the image size is beyond what Python allows to read. It can not be concluded whether this is a malicious file.
6. Not all the JPEG files contain the same set of tags.

3.4.1 Strategy I: (using the length of the tags as features)

Once the tags are extracted from the JPEG files, we create a unique list of tags found across all the JPEG files. A data frame is formed with each of these tags as columns. The length of each tag in every JPEG file is determined. If a tag is not available in a JPEG File, the corresponding column in the data frame is assigned a value of zero. So, now there is a data frame containing only numbers. A Logistic Regression model is developed using this data frame. The accuracy obtained through this model is 56%. A Random Forest model is trained using the same data frame. The Random Forest model performed slightly better with an accuracy of 58%. Also, we prepare an artificial neural network using Tensor Flow API, and the model gives an accuracy of about 72%.

3.4.2 Strategy II: (forming TF-IDF)

In Strategy II, we have decided to use TF-IDF (term frequency-inverse document frequency) over the tags. A string is formed by concatenating all the tags available in each JPEG file then TF-IDF is created from these strings. After making the TF-IDF, the Decision Tree model is trained. This model gives an accuracy of 98.9% during training. This result encourages a Random Forest model to be created using the same TF-IDF. Both models' performance results are the same. This work only contains the model developed using strategy II.

3.5 Working of the System

The model must deploy on any server to which the web server has access. The model may reside on a server different from the web server because the model may need to be rebuilt from time to time. Now, rebuilding the model requires a machine with more RAM and virtual memory. The model for this project is built on a device having 64 GB RAM and a 2 TB Hard Disk.

The system starts when the web server is started. The web server first loads the model and then listens on a port for the Client. One should always keep the web server running so that the Client can contact it. For this to happen, the web server is started using the **nohup** command. **nohup** command ensures that the web server keeps running even if the user who started the web server logs out of the system. Once the web server is up and running, the Client Application can call the web server's IP and port and then send files for evaluation.

3.6 Building the Model

In this section, we discuss the code used to build the model.

3.6.1 Constants used

The following constants are used in the code. Constants used in the code are as follows -

```
1 # Constants
2 FILE_NAME_COLUMN_NAME = 'FileName'
```

```
3 FILE_TYPE_COLUMN_NAME =  'FileType'
4 TAG_STRING_COLUMN_NAME =  'TagString'
5 NUMERIC_COLUMN_IDENTIFIER = 'AAA'
6 BENIGN_FILE = 0
7 FILE_WITH_MALWARE = 1
```

3.6.2 Functions used to extract EXIF tags from JPEG files

The following three functions are used to extract the EXIF tags from the JPEG files.

extractTagsFromADirectory: This function takes a single file name or a set of file names and returns the tags extracted from all the JPEG files as a List. Along with the tags found in the set of JPEG files, this function returns the number of valid JPEG files, number of invalid JPEG files and the number of JPEG files which contained no tags. This function calls the functions JPEGFileFeatureExtractorToDictionary() and isImageFile().

```
1  import glob
2
3  def extractTagsFromADirectory(inputDirectory):
4  # Declare Counters
5  numberOfValidFiles = 0
6  numberOfInvalidFiles = 0
7  numberOfFilesWithoutTags = 0
8
9  # Create an Empty List to hold all the features of all the
       files
10 returnValue = []
11
12
13 # Loop through all the files in the Input Directory
14 for file in glob.glob(inputDirectory):
15 # Create an empty Dictionary
16 oneFileFeatures = {}
17
18 try:
19 # Read the file and extract the features
20 fileFeatures = JPEGFileFeatureExtractorToDictionary(file)
21
22 # If the File had some features, then create an entry for
       the file
23 if len(fileFeatures.keys()) > 0:
24 # Write the File Name
25 oneFileFeatures[FILE_NAME_COLUMN_NAME] = file
```

```
26
27 # Add the File Features to the main Dictionary
28 oneFileFeatures.update(fileFeatures)
29
30 # Add the entry to the return value
31 returnValue.append(oneFileFeatures)
32 numberOfValidFiles = numberOfValidFiles + 1
33 else:
34 #Check if the file is a valid Image File
35 if isImageFile(file):
36 numberOfFilesWithoutTags = numberOfFilesWithoutTags + 1
37 else:
38 numberOfInvalidFiles = numberOfInvalidFiles + 1
39 except:
40 # Check if the file is a valid Image File
41 if isImageFile(file):
42 numberOfFilesWithoutTags = numberOfFilesWithoutTags + 1
43 else:
44 numberOfInvalidFiles = numberOfInvalidFiles + 1
45
46 return (returnValue, numberOfValidFiles,
          numberOfInvalidFiles, numberOfFilesWithoutTags)
```

JPEGFileFeatureExtractorToDictionary: This function takes one JPEG file as input and returns all the EXIF tags in the JPEG file in a Dictionary as output.

```
1 from PIL import Image
2 from PIL.ExifTags import TAGS
3
4 def JPEGFileFeatureExtractorToDictionary(imageFile):
5 #Declare an empty Dictionary
6 returnValue = {}
7
8 # Read the image data using PIL
9 image = Image.open(imageFile)
10
11 # Extract EXIF data
12 exifdata = image.getexif()
13
14 # Iterating over all EXIF data fields
15 for tag_id in exifdata:
16 # Get the tag name and the associated data
17 tag = TAGS.get(tag_id, tag_id)
18 data = exifdata.get(tag_id)
19
```

```
20 # Decode bytes
21 if isinstance(data, bytes):
22 data = data.decode('iso8859-1')
23
24 returnValue[tag] = data
25
26 return returnValue
```

isImageFile: This function takes a file as an input and returns TRUE if the file is a JPEG file and returns FALSE if the file is not a JPEG file.

```
1 from PIL import Image
2
3 def isImageFile(imageFileName):
4 returnValue = True
5
6 try:
7 img = Image.open('./' + imageFileName) # open the image file
8 img.verify() # verify that it is an image
9
10 except (IOError, SyntaxError) as e:
11 returnValue = False
12
13 return returnValue
```

Using these functions, the features from the benign JPEG files and the JPEG files containing malware are extracted as shown in Listing 3.2.

```
1 #Code for Extracting the features from the JPEG Files
2
3 benignFileFeatures, numValidFiles, numInvalidFiles,
   numFilesWithoutTags = extractTagsFromADirectory("./Data/
   clean_jpeg/*.j*")
4 print("Valid JPEG Files = %d\nInvalid Image Files = %d\nJPEG
   Files without Tags = %d" % (numValidFiles,
   numInvalidFiles, numFilesWithoutTags))
5
6 Valid JPEG Files = 4582
7 Invalid Image Files = 0
8 JPEG Files without Tags = 816
9
10 malwareFileFeatures, numValidFiles, numInvalidFiles,
   numFilesWithoutTags = extractTagsFromADirectory("./Data/
   malicious_files/*")
11 print("Valid JPEG Files = %d\nInvalid Image Files = %d\nJPEG
   Files without Tags = %d" % (numValidFiles,
   numInvalidFiles, numFilesWithoutTags))
```

```
12
13  Valid JPEG Files = 400
14  Invalid Image Files = 12
15  JPEG Files without Tags = 3
```

Listing 3.2   Feature extraction for malicious and benign files.

3.6.3 Example of EXIF tags

Listing 3.3 shows an example of EXIF tags available in JPEG files.

```
1   'GPSInfo': 1108,
2   'ResolutionUnit': 2,
3   'ExifOffset': 204,
4   'Make': 'Apple',
5   'Model': 'iPhone 4S',
6   'Software': '9.3.5',
7   'Orientation': 1,
8   'DateTime': '2017:09:07 16:03:42',
9   'YCbCrPositioning': 1,
10  'XResolution': 72.0,
11  'YResolution': 72.0
```

Listing 3.3   EXIF tags example.

3.6.4 Unique keys extraction for all files

From the last step, we have two dictionaries containing all the tags extracted from all the clean JPEG files and from all the JPEG files containing malware. Now, a list of all the unique tags is prepared. Another significant aspect noticed is that there are some tags which are numeric. Since a data frame can not have a column name containing only numbers, these tags are prefixed with a fixed string. The list of unique tags extraction code is given in Listing 3.4.

```
1   featureList = set()
2
3   for i in benignFileFeatures:
4   for k in i.keys():
5   if type(k) == int:
6   featureList.add(NUMERIC_COLUMN_IDENTIFIER + str(k))
7   else:
8   featureList.add(k)
9
10  for i in malwareFileFeatures:
11  for k in i.keys():
```

```
12  if type(k) == int:
13  featureList.add(NUMERIC_COLUMN_IDENTIFIER + str(k))
14  else:
15  featureList.add(k)
```

Listing 3.4   unique tags extraction.

Once the feature list is prepared, it can be viewed by viewing the set featureList. The important aspect is that this feature list is required when the features are extracted from the test dataset and/or from the dataset of the JPEG files which need evaluating when the system is in production.

3.6.5 Preparation of data frame for creating TF-IDF

To form the TF-IDF, we need a string for each JPEG file containing the details of the extracted EXIF tags. However, we need to only consider the features that we have shortlisted in the Section 3.6.2. To be able to do this, we use the following function and apply it on the benign files dataset and the malicious files dataset.

```
1  #Function fillDataInDataFrame()
2  import pandas as pd
3
4  def fillDataInDataFrame(featureDictionary, fileType):
5  # Create an Empty DataFrame object
6  df = pd.DataFrame()
7
8  for record in featureDictionary:
9  # Create an empty Dictionary
10  oneRecord = {}
11
12  # Create an empty string
13  recordString = ""
14
15  # Loop through all the features in a record
16  for k in record.keys():
17  # Extract the value for the key and append to the record
        string
18  # This will be used for TFID
19  # Add a SPACE between each tag value
20  # Do not include File Name
21  if k != FILE_NAME_COLUMN_NAME:
22  if k in featureList:
23  recordString = recordString + str(record[k]) + " "
24
```

```
25 # Add the record string as a separate column in the record
26 oneRecord [TAG_STRING_COLUMN_NAME] = recordString [:-1]
27
28 # Add column to mark Dependent Column as File Type
29 oneRecord [FILE_TYPE_COLUMN_NAME] = pd.to_numeric(fileType,
        downcast='integer')
30
31 # Add the Record to the Data Frame
32 df = df.append(oneRecord, ignore_index=True)
33
34 return df
```

This function can be used to prepare the data frame to contain all the features extracted from all the JPEG files in the training dataset as shown in Listing 3.5. Each record of the data frame will be labeled as Benign, or Malware based on the file from which the features are extracted.

```
1 #Code to prepare the data frame of classified features
2 import pandas as pd
3
4 benignFileDF = fillDataInDataFrame(benignFileFeatures,
5 BENIGN_FILE)
6
7 malignantFileDF = fillDataInDataFrame(malwareFileFeatures,
8 FILE_WITH_MALWARE)
9
10 df = pd.concat([benignFileDF, malignantFileDF], ignore_index
        =True)
11
12 df[FILE_TYPE_COLUMN_NAME] = pd.to_numeric(df[
        FILE_TYPE_COLUMN_NAME], downcast='integer')
```

Listing 3.5 Data frame construction using all JPEG files.

The next step is to eliminate the $CHR(0)$ from the records from the overall data frame. This is because if the program encounters a $CHR(0)$, the reading of the record will terminate at that point.

```
1 # Code to remove CHR(0)
2 dfTFID = df [[TAG_STRING_COLUMN_NAME, FILE_TYPE_COLUMN_NAME
        ]].copy()
3
4 dfTFIDClean = pd.DataFrame()
5 for i in dfTFID.index:
6 dfTFIDClean.loc[i, TAG_STRING_COLUMN_NAME] = dfTFID.iloc[i,
        0].replace(chr(0), '')
```

```
7  dfTFIDClean.loc[i, FILE_TYPE_COLUMN_NAME] = dfTFID.iloc[i,
       1]
8
9  dfTFIDClean[FILE_TYPE_COLUMN_NAME].value_counts()
10
11  0  4582
12  1  400
13  Name: FileType, dtype: int64
```

3.6.6 Forming the TF-IDF

TF-IDF (term frequency-inverse document frequency) is a statistical measure that evaluates how relevant a word is to a document in a collection of documents. It is done by multiplying two metrics – how many times a word appears in a document and the inverse document frequency of a word across a set of documents. In this case, each JPEG file is a document in the collection, and each tag in each JPEG file is a term.

You notice that there are 64,774 features in our dataset after forming the TF-IDF. The TF-IDF is stored in an array with each element of the array corresponding to a JPEG file in the same order as we formed the data frame. It creates the set of independent variables for our machine learning model. The independent variables are stored in the variable X. The dependent variable is the FILE-TYPE set as recorded while loading the JPEG files (remember that the clean JPEG files are stored in a separate directory, and all these files are read as a set and marked as benign. Similarly, all the JPEG files with malware are stored in a different directory and read as a set and marked as MALWARE).

```
1  #Code to form the TF-IDF
2  from sklearn.feature_extraction.text import TfidfVectorizer
3
4  tfidfconverter = TfidfVectorizer(max_features=90000, min_df
       =1, max_df=0.7)
5  X = tfidfconverter.fit_transform(dfTFIDClean.TagString).
       toarray()
6
7  y = dfTFID.FileType
8
9  # Save the Decision Tree Model to a file
10  import pickle
11
12  pickle.dump(tfidfconverter, open("./TFIDFConverter", 'wb'))
13
```

```
14 X.shape
15
16 (4982, 64774)
17
18 y.value_counts()
19
20 0  4582
21 1  400
22 Name: FileType, dtype: int64
```

Note that we need to save the TF-IDF converter object to a file. This object will be used by the web server to create the TF-IDF for the new files which have to be evaluated by the model.

3.6.7 Handling the imbalanced datasets

We have 4582 clean JPEG files and 400 JPEG files with malware. So, our dataset is imbalanced as we have more than 10 times the number of clean JPEG files compared to the number of JPEG files with malware. Imbalanced datasets will cause algorithms like Random Forest and Decision Tree (and most other algorithms) not to function properly. We can understand this specifically in the case of the Random Forest algorithm. We know that in the Random Forest algorithm, the datasets are split both vertically and horizontally at random [9]. A separate decision tree model then analyzes each split. When we have smaller data points for one set of data, one or more of the horizontal splits may get data from only one class. It will cause those decision trees not to be able to see both types of data. Thus, the classifications will not be proper.

To resolve the problem of the imbalanced dataset, we oversample the dataset. Oversampling means that we try to increase the number of data points in the class, which has a smaller number of data points similar to the number of data points in the other set with more data points. For oversampling, SMOTE (Synthetic Minority Oversampling Technique) algorithm is used.

```
1 #Code to over-sample using SMOTE
2 import imblearn as ib
3
4 oversample = ib.over_sampling.SMOTE()
5 X, y = oversample.fit_resample(X, y)
6
7 X.shape
8
9 (9164, 64774)
```

```
10
11 y.value_counts()
12
13 0  4582
14 1  4582
15 Name: FileType, dtype: int64
```

Notice that after applying SMOTE, we have 9164 data points in our dataset and there are 4582 data points for clean JPEG files and 4582 data points for JPEG files with malware.

3.6.8 Development of decision tree model

We implement the model using the Decision Tree algorithm as shown in Listing 3.6.

```
1  #Code to form the classification model using Decision Tree
      Algorithm
2  from sklearn.model_selection import cross_val_score
3  from sklearn.model_selection import RepeatedStratifiedKFold
4  from sklearn.tree import DecisionTreeClassifier
5
6  # Instantiate the Decision Tree Model
7  modelDT = DecisionTreeClassifier()
8
9  # Evaluate the model
10 cv = RepeatedStratifiedKFold(n_splits=10, n_repeats=3,
      random_state=1)
11 scores = cross_val_score(modelDT, X, y, scoring='roc_auc',
      cv=cv, n_jobs=-1)
12 print('Mean ROC AUC: %.5f' % scores.mean())
13
14 Mean ROC AUC: 0.99347
15
16 # Create the Decision Tree Model
17 modelDT.fit(X, y)
18
19 # Save the Decision Tree Model to a file
20 import pickle
21
22 pickle.dump(modelDT, open("./DTModelMalwareDetection", 'wb')
      )
```

Listing 3.6   Decision Tree model.

Notice that the model has an accuracy of 99.347% on the training data when we generate the cross-validation score. Now our model is ready, let us test the model on the training dataset. Given Listing 3.7 shows the code for making prediction using the model on the training dataset. Figure 3.1 and 3.2 show the corresponding confusion matrices generated using training and testing data, respectively.

```python
#Code to make predictions on the training data set using the
    model created using Decision Tree Algorithm

from sklearn import metrics
import matplotlib.pyplot as plt
import seaborn as sns

# Make the predictions
y_pred = modelDT.predict(X)

# Generate the Confusion Matrix
cm = metrics.confusion_matrix(y, y_pred)

# Plot the Confusion Matrix
ax = plt.subplot()
sns.heatmap(cm, annot=True, fmt='g', ax=ax);

ax.set_xlabel('Predicted labels');
```

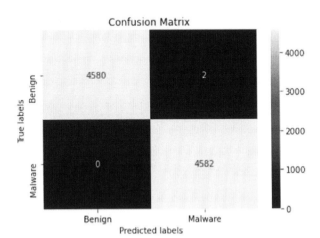

Figure 3.1 Confusion matrix of predictions on training data using model developed using Decision Tree algorithm.

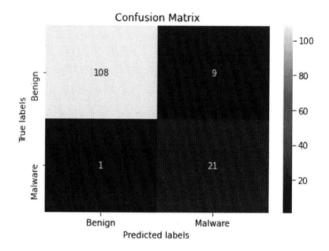

Figure 3.2   Confusion matrix of predictions on test data using model developed using Decision Tree algorithm.

```
18  ax.set_ylabel('True labels');
19  ax.set_title('Confusion Matrix');
20  ax.xaxis.set_ticklabels(['Benign', 'Malware']);
21  ax.yaxis.set_ticklabels(['Benign', 'Malware']);
22
23  print("\n\nConfusion Classification Report\n")
24  print(metrics.classification_report(y, y_pred))
25  Confusion Classification Report
26
27                precision    recall  f1-score    support
28  0                  1.00      1.00      1.00       4582
29  1                  1.00      1.00      1.00       4582
30  accuracy                               1.00       9164
31  macroavg           1.00      1.00      1.00       9164
32  weighted avg       1.00      1.00      1.00       9164
```

Listing 3.7   Decision Tree model prediction.

Now, we test the model on the test dataset. Before we make predictions on the test dataset, we construct the data from the test dataset suitable for the model. Listing 3.8 shows the necessary code to prepare the test dataset before predictions can be made for the same.

Now we have the test dataset, which can be used to make predictions using the Decision Tree model. Listing 3.9 states the mechanism to make predictions on the test dataset and check the results. We see that we get an accuracy of 92.805% on the test dataset.

```
1  #Code to prepare the test data set before predictions can be
       made on them
2  # Read the data for the Benign Files
3
4
5
6  testCleanFeatures, numValidFiles, numInvalidFiles,
       numFilesWithoutTags = extractTagsFromADirectory("./Data/
       ValidationSet-Clean/*.j*")
7  print("\nBENIGN FILES\n------------\nValid JPEG Files = %d\
       nInvalid Image Files = %d\nJPEG Files without Tags = %d"
       % (numValidFiles, numInvalidFiles, numFilesWithoutTags)
       )
8
9  BENIGN FILES
10 ------------
11 Valid JPEG Files = 117
12 Invalid Image Files = 0
13 JPEG Files without Tags = 175
14
15
16 # Form the data frame
17 testCleanDF = fillDataInDataFrame(testCleanFeatures,
       BENIGN_FILE)
18
19 # Read the data for the file with Malware
20 testMalwareFeatures, numValidFiles, numInvalidFiles,
       numFilesWithoutTags = extractTagsFromADirectory("./Data/
       ValidationSet-Malicious/*")
21 print("\nMALWARE FILES\n-------------\nValid JPEG Files = %d
       \nInvalid Image Files = %d\nJPEG Files without Tags = %d
       " % (numValidFiles, numInvalidFiles, numFilesWithoutTags
       ))
22
23 MALWARE FILES
24 -------------
25 Valid JPEG Files = 22
26 Invalid Image Files = 0
27 JPEG Files without Tags = 0
28
29
30 # Form the data frame
```

```
31 testMalwareDF = fillDataInDataFrame(testMalwareFeatures,
      FILE_WITH_MALWARE)
32
33 # Combine the 2 data frames formed above
34 testDF = pd.concat([testCleanDF, testMalwareDF],
      ignore_index=True)
35 testDF[FILE_TYPE_COLUMN_NAME] = pd.to_numeric(testDF[
      FILE_TYPE_COLUMN_NAME], downcast='integer')
36
37 # Create data for TFID
38 testTFIDDF = testDF[[TAG_STRING_COLUMN_NAME,
      FILE_TYPE_COLUMN_NAME]].copy()
39
40 # Clean the data
41 testTFIDClean = pd.DataFrame()
42 for i in testTFIDDF.index:
43 testTFIDClean.loc[i, TAG_STRING_COLUMN_NAME] = testTFIDDF.
      iloc[i, 0].replace(chr(0), ' ')
44 testTFIDClean.loc[i, FILE_TYPE_COLUMN_NAME] = testTFIDDF.
      iloc[i, 1]
45
46 # Drop NULL Values
47 testTFIDClean = testTFIDClean.dropna()
48 testTFIDClean[FILE_TYPE_COLUMN_NAME]= pd.to_numeric(
      testTFIDClean[FILE_TYPE_COLUMN_NAME], downcast=
49      'integer')
50 testTFIDClean[FILE_TYPE_COLUMN_NAME].value_counts()
51
52 0  117
53 1  22
54 Name: FileType, dtype: int64
55
56
57 # Create the TFID
58 X_test = tfidfconverter.transform(testTFIDClean.TagString).
      toarray()
59 y_test = testTFIDClean.FileType
```

Listing 3.8   Prepare test data.

```
1 #Code to make predictions on the test data and evaluate the
      results
2
3 # Make the predictions
4 y_pred_test = modelDT.predict(X_test)
5
```

```
 6  # Generate the Confusion Matrix
 7  cm = metrics.confusion_matrix(y_test, y_pred_test)
 8
 9  # Plot the Confusion Matrix
10  ax = plt.subplot()
11  sns.heatmap(cm, annot=True, fmt='g', ax=ax);
12
13  ax.set_xlabel('Predicted labels');
14  ax.set_ylabel('True labels');
15  ax.set_title('Confusion Matrix');
16  ax.xaxis.set_ticklabels(['Benign', 'Malware']);
17  ax.yaxis.set_ticklabels(['Benign', 'Malware']);
```

Listing 3.9 Prediction on test data.

3.6.9 Development of random forest model

Training the model using Random Forest algorithm is like creating the model using the Decision Tree algorithm. We use the data that we already prepared in the Section 3.6.6. Listing 3.10 shows the code for the classification model generation using Random Forest classifier.

```
 1  #Code to form the classification model using Random Forest
        Algorithm
 2  from sklearn.model_selection import cross_val_score
 3  from sklearn.model_selection import RepeatedStratifiedKFold
 4  from sklearn.ensemble import RandomForestClassifier
 5
 6  # Instantiate the Decision Tree Model
 7  modelRF = RandomForestClassifier()
 8
 9  # Evaluate the model
10  cv = RepeatedStratifiedKFold(n_splits=10, n_repeats=3,
        random_state=1)
11  scores = cross_val_score(modelRF, X, y, scoring='roc_auc',
        cv=cv, n_jobs=-1)
12  print('Mean ROC AUC: %.5f' % scores.mean())
13
14  Mean ROC AUC: 0.99995
15
16  # Create the Random Forest Model
17  modelRF.fit(X, y)
18
19  # Save the Random Forest Model to a file
20  import pickle
```

```
21
22 pickle.dump(modelRF, open("./RFModelMalwareDetection", 'wb')
      )
```

Listing 3.10   Classification using Random Forest model.

Notice that the model has an accuracy of 99.995% on the training data when we generate the cross-validation score. We now test this model on the training dataset as shown in Listing 3.11.

```
1  #Code to make predictions on the training data set using the
      model created using the Random Forest Algorithm
2  from sklearn import metrics
3  import matplotlib.pyplot as plt
4  import seaborn as sns
5
6  # Make the predictions
7  y_pred = modelRF.predict(X)
8
9  # Generate the Confusion Matrix
10 cm = metrics.confusion_matrix(y, y_pred)
11
12 # Plot the Confusion Matrix
13 ax = plt.subplot()
14 sns.heatmap(cm, annot=True, fmt='g', ax=ax);
15
16 ax.set_xlabel('Predicted labels');
17 ax.set_ylabel('True labels');
18 ax.set_title('Confusion Matrix');
19 ax.xaxis.set_ticklabels(['Benign', 'Malware']);
20 ax.yaxis.set_ticklabels(['Benign', 'Malware']);
21
22 print("\n\nConfusion Classification Report\n")
23 print(metrics.classification_report(y, y_pred))
24
25 Confusion Classification Report
26
27               precision    recall  f1-score    support
28 0               1.00        1.00     1.00        4582
29 1               1.00        1.00     1.00        4582
30 accuracy                             1.00        9164
31 macroavg        1.00        1.00     1.00        9164
32 weighted avg    1.00        1.00     1.00        9164
```

Listing 3.11   Random Forest model prediction using training data.

We can use the same test data we prepared in Section 3.6.7 to test our Random Forest model. The code shown in Listing 3.12 states the mechanism

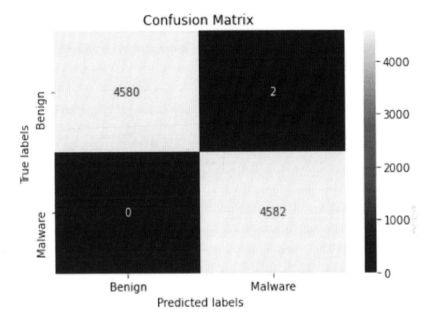

Figure 3.3   Confusion Matrix of Predictions on Training Data using model developed using Random Forest algorithm

to make predictions on the test dataset using the model created using Random Forest algorithm and check the results. We see that we get an accuracy of 92.805% on the test dataset. Figure 3.3 and3.4 show the corresponding confusion matrices generated using training and testing data.

```
1 #Code to make predictions on the test data and evaluate the
     results
2 # Make the predictions
3 y_pred_test = modelRF.predict(X_test)
4
5 # Generate the Confusion Matrix
6 cm = metrics.confusion_matrix(y_test, y_pred_test)
7
8 # Plot the Confusion Matrix
9 ax = plt.subplot()
10 sns.heatmap(cm, annot=True, fmt='g', ax=ax);
11
12 ax.set_xlabel('Predicted labels');
13 ax.set_ylabel('True labels');
14 ax.set_title('Confusion Matrix');
```

```
15 ax.xaxis.set_ticklabels(['Benign', 'Malware']);
16 ax.yaxis.set_ticklabels(['Benign', 'Malware']);
```

Listing 3.12   Random Forest model prediction using testing data.

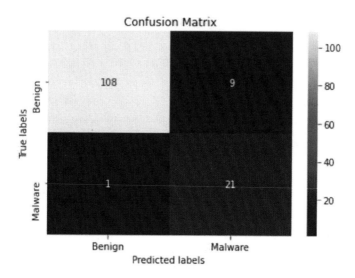

Figure 3.4   Confusion matrix of predictions on test data using model developed using Random Forest algorithm.

3.7 Conclusions on the Model

We see that both the models developed demonstrate a similar accuracy in training and testing. The significant aspect of the model is that the number of false positive are either missing or very minimal. So, the model takes away maximum risk from the users of the model. However, we notice that we get more than 99% accuracy on the training dataset. The accuracy drops of about 92% on the test dataset. So, we can conclude that the model suffers from overfitting. The overfitting can be attributed to the complexity of the model. So, the model has a high variance.

We can solve the overfitting problem by conducting a principal component analysis (PCA) on the extracted features before forming TF-IDF. This will help in simplifying the model and thus the variance will reduce.

3.8 Creating the Web Service

Now we have our model, we can develop the web service through which the model can be accessed by any client. Creating a web service will allow any kind of client application to be able to use the model. Listing 3.13 shows the code for the web service. The web service takes a file as an input. The web service evaluates whether the file is a valid JPEG file. For every valid JPEG file, the web service reports whether the file contains virus or is a clean JPEG file.

```
1  #Code for the Web Service
2  #!/usr/bin/env python
3  # -*- coding: UTF-8 -*-
4
5  import pandas as pd
6  import os
7  import cherrypy
8  import pickle
9  from PIL import Image
10 from PIL.ExifTags import TAGS
11 import glob
12
13 config = {
14     'global' : {
15         'server.socket_host' : '127.0.0.1',
16         'server.socket_port' : 8080,
17         'server.thread_pool' : 8,
18         'server.max_request_body_size' : 0,
19         'server.socket_timeout' : 60
20     }
21 }
22
23 class App:
24
25     def __init__(self):
26         self.RFmodel = pickle.load(open
27 ('RFModelMalwareDetection', 'rb'))
28         self.DTmodel = pickle.load(open
29 ('DTModelMalwareDetection', 'rb'))
30         self.TFTDFconverter = pickle.load(open
31 ('TFIDFConverter', 'rb'))
32         self.featureList = pickle.load(open('FeatureList',
    'rb'))
33         self.FILE_NAME_COLUMN_NAME = 'FileName'
34         self.FILE_TYPE_COLUMN_NAME = 'FileType'
```

```python
35          self.TAG_STRING_COLUMN_NAME = 'TagString'
36          self.NUMERIC_COLUMN_IDENTIFIER = 'AAA'
37
38
39      def isImageFile(self, imageFileName):
40          returnValue = True
41
42          try:
43              img = Image.open(imageFileName) # open the image
        file
44              img.verify() # verify that it is an image
45          except (IOError, SyntaxError) as e:
46              returnValue = False
47
48          return returnValue
49
50      def extractTagsFromAFile(self, inputFile):
51          tagsAvailable = True
52
53          # Create an Empty List to hold all the features
54          returnValue = []
55
56          # Create an empty Dictionary
57          oneFileFeatures = {}
58
59          try:
60              # Read the file and extract the features
61              fileFeatures = self.
        JPEGFileFeatureExtractorToDictionary(inputFile)
62
63              # If the File had some features, then create an
        entry for the file
64              if len(fileFeatures.keys()) > 0:
65                  # Add the File Features to the main
        Dictionary
66                  oneFileFeatures.update(fileFeatures)
67
68                  # Add the entry to the return value
69                  returnValue.append(oneFileFeatures)
70              else:
71                  tagsAvailable = False
72
73          except:
74              tagsAvailable = False
75
76          return (returnValue, tagsAvailable)
```

```
77
78    def  JPEGFileFeatureExtractorToDictionary(self ,  imageFile
      ) :
79        #Declare  an  empty  Dictionary
80        returnValue  =  {}
81
82        try :
83            # read  the  image  data  using  PIL
84            image  =  Image.open(imageFile)
85
86            # extract  EXIF  data
87            exifdata  =  image.getexif()
88
89            # iterating  over  all  EXIF  data  fields
90            for  tag_id  in  exifdata :
91                # get  the  tag  name
92                tag  =  TAGS.get(tag_id ,  tag_id)
93                data  =  exifdata.get(tag_id)
94                # decode  bytes
95                if  isinstance(data ,  bytes) :
96                    data  =  data.decode('iso8859-1')
97
98                #print(tag)
99                returnValue[tag]  =  data
100       except :
101           pass
102
103       return  returnValue
104
105   def  fillDataInDataFrame(self ,  featureDictionary) :
106       # Create  an  Empty  DataFrame  object
107       df  =  pd.DataFrame()
108
109       for  record  in  featureDictionary :
110           # Create  an  empty  Dictionary
111           oneRecord  =  {}
112
113           # Create  an  empty  string
114           recordString  =  ""
115
116           # Initialise  all  the  columns
117           for  colName  in  self.featureList :
118               oneRecord[colName]  =  pd.to_numeric(0 ,
      downcast='integer')
119
120           # Loop  through  all  the  features  in  a  record
```

```
121            for k in record.keys():
122                    # Extract the value for the key and append
    to the record string
123                    # This will be used for TFID
124                    # Add a SPACE between each tag value
125                    # Do not include File Name
126                    if k != self.FILE_NAME_COLUMN_NAME:
127 if k in self.featureList:
128                        recordString = recordString + str(record
    [k]) + " "
129
130                # Add the record string as a separate column in
    the record
131                oneRecord[self.TAG_STRING_COLUMN_NAME] =
    recordString[:-1]
132
133                # Add column to mark Dependent Column as Benign
    File
134                oneRecord[self.FILE_TYPE_COLUMN_NAME] = pd.
    to_numeric(0, downcast='integer')
135
136                # Add the Record to the Data Frame
137                df = df.append(oneRecord, ignore_index=True)
138
139        return df
140
141    def prepareFileForAnalysis(self, imageFileName):
142        returnValue = 0
143        X = None
144
145        testFeatures, tagsAvailable = self.
    extractTagsFromAFile(imageFileName)
146        if tagsAvailable == False:
147            returnValue = -1
148        else:
149            testDF = self.fillDataInDataFrame(testFeatures)
150            testDF[self.FILE_TYPE_COLUMN_NAME] = pd.
    to_numeric(testDF[self.FILE_TYPE_COLUMN_NAME], downcast=
    'integer')
151            testTFIDDF = testDF[[self.TAG_STRING_COLUMN_NAME
    , self.FILE_TYPE_COLUMN_NAME]].copy()
152
153            testTFIDClean = pd.DataFrame()
154            for i in testTFIDDF.index:
```

```
155        testTFIDClean.loc[i, self.
    TAG_STRING_COLUMN_NAME] = testTFIDDF.iloc[i, 0].replace(
    chr(0), '')
156        testTFIDClean.loc[i, self.
    FILE_TYPE_COLUMN_NAME] = testTFIDDF.iloc[i, 1]
157
158        testTFIDClean = testTFIDClean.dropna()
159        testTFIDClean[self.FILE_TYPE_COLUMN_NAME]= pd.
    to_numeric(testTFIDClean[self.FILE_TYPE_COLUMN_NAME],
    downcast='integer')
160
161        X = self.TFTDFconverter.transform(testTFIDClean.
    TagString).toarray()
162
163    return X, returnValue
164
165 @cherrypy.expose
166 def upload(self, ufile):
167     upload_path = os.path.normpath('./data/')
168     upload_file = os.path.join(upload_path, ufile.
    filename)
169     size = 0
170
171     returnValue = 0
172
173     with open(upload_file, 'wb') as out:
174         while True:
175             data = ufile.file.read(8192)
176             if not data:
177                 break
178             out.write(data)
179             size += len(data)
180
181     # Check whether File is an Image File
182     if(self.isImageFile(upload_file) == False):
183         returnValue = -1 # Not a JPEG File
184     else:
185         X, fileStatus = self.prepareFileForAnalysis(
    upload_file)
186         if fileStatus == -1:
187             returnValue = -2 # JPEG File does not have
    any tags
188         else:
189             returnValue = self.DTmodel.predict(X)[0]
190
191     out = '''
```

```
192                    returnValue: {}
193                    length: {}
194                    filename: {}
195                    mime-type: {}
196                    ''' .format(returnValue, size, ufile.
        filename, ufile.content_type, data)
197
198            return out
199
200
201  if __name__ == '__main__':
202      cherrypy.quickstart(App(), '/', config)
```

Listing 3.13   Webservice code.

3.9 Creating a Simple Client Application

Listing 3.14 shows the code for a simple client application which can test one JPEG file at a time using the web service created in Section 3.8.

```
1  #Code for a simple Client Application
2  #!/usr/bin/env python
3  # -*- coding: UTF-8 -*-
4
5  import requests
6  import glob
7
8  url = 'http://127.0.0.1:8080/upload'
9
10 for file in glob.glob('./Data/ValidationSet-Clean/*.j*'):
11     files = {'ufile': open(file, 'rb')}
12     r = requests.post(url, files=files, verify = False)
13     print(r)
14     print(r.text)
15
16 for file in glob.glob('./Data/ValidationSet-Malicious/*'):
17     files = {'ufile': open(file, 'rb')}
18     r = requests.post(url, files=files)
19     print(r)
20     print(r.text)
21 }
```

Listing 3.14   Simple client application code.

3.10 Sample of a Sophisticated Client Application

Figure 3.5 shows the user interface of a sophisticated client application which calls the web service mentioned in Section 3.19. This client application resides on a Windows machine. The application watches a particular directory on the machine. Whenever a new file is created or modified on the directory being watched, this client application picks up the file and sends to the web service for evaluation. If the web service reports that input file contains malware, the client application moves the file to a directory set for quarantine.

3.11 Detecting Malware in ELF Files

Malware can be defined as any malicious program that intends to cause damage to a computer, server, client, or computer network. Malware [81] can be categorized and classified into different types, depending on its functionality. The five malware categories classified by the proposed model are listed below:

- **Backdoor:** Malicious software capable of opening unauthorized access to remote systems.
- **Botnet:** A botnet is a network of compromised systems that allow a hacker to control them remotely and execute commands.
- **DDOS:** A DDOS malware can launch distributed denial-of-service attacks by bombarding victim machine(s) with unnecessary packets and thus rendering the machines unresponsive to legitimate requests.

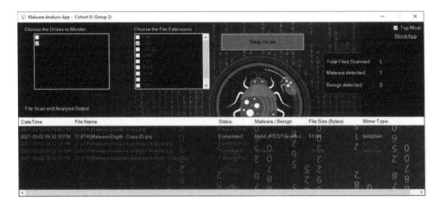

Figure 3.5 Client application user interface.

- **Trojan:** A trojan is any kind of malware which disguises itself as something legitimate and misleads users of its true intent.
- **Virus:** Malicious code capable of replicating itself by inserting its own code into other programs is termed as a virus.

3.12 About ELF Files

First, we need to discuss the structure of the ELF files. This is because the project starts with extracting information from the ELF files. Without this information, it will be impossible to build any kind of model for classifying the files as good or malware [80]. The information in the ELF files can be broadly classified into three types of data: data in the File Header, data in the Program Header, and data in the Sections. The ELF files have one File Header, one Program Header, and many Sections of information. Figure 3.6 shows the structure of the ELF files.

3.12.1 ELF file header

Every ELF file has the first 2 bytes as 0x7F. By checking these 2 bytes, we can determine whether a file is a valid ELF file or not. The ELF file Header defines whether to use 32- or 64-bit addresses. The header contains three fields that are affected by this setting and offset other fields that follow them. The ELF header is 52 or 64 bytes long for 32-bit and 64-bit binaries respectively. The ELF file

Structure of the ELF Files

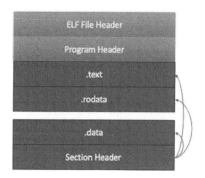

The illustration on the left shows the structure of the ELF Files in the memory of a machine when an ELF File is loaded for execution.

Every ELF File has only 1 File Header and only 1 Program Header. However, an ELF File can have many sections and each section has a Section Header.

.text is the program text of the program.

.rodata is the Read Only data of the program.

.data are the sections where all the transient data of the program are loaded.

Figure 3.6 Structure of the ELF files.

Header contains information on the target operating system. This information is stored in 1-byte. The target operating system can be any flavour of Unix like Solaris, HP-UX, NetBSD, Linux, etc. There is also a byte in the ELF file Header which states the version of the target operating system.There is a 2-bytes information on the type of ELF file in the ELF file Header. The type of ELF file can be an executable,dynamic linked library, Core dump, etc.There is a 2-byte information on the type of machine for which the ELF file has been built. The information can be SPARC, Intel 80860, NEX, Power PC, Digital Alpha, Hitachi, etc.After all this information, there is 4-byte pointer to where the process should start execution. This is a very vulnerable point in the ELF file. The malware designers try to alter this pointer to point the program to the memory location where the malware developers induct their code.Then, there is a 4-byte pointer to the Program Header. This is also a vulnerable point in the ELF files for the same reason discussed above.Following this, there is a 4-byte pointer to the Section Header table. This is also a vulnerable point in the ELF files for the same reason discussed above.Following this is a 4-byte pointer to the Section Header table. Then, there is a 2-byte pointer to the size of the ELF file Header. This value needs to be 64 bytes for 64-bit format and 52 bytes for the 32-bit format.There is some more information like the number of entries in the Program Header, the size of the Section Header, etc.

3.12.2 Program header

The program header table tells the system how to create a process image. It is found at file offset e_phoff, and consists of e_phnum entries, each with size e_phentsize. The layout is slightly different in 32-bit ELF vs 64-bit ELF, because the p_flags are in a different structure location for alignment reasons.The first information in the Program Header is the type of the segment-whether the segment is a loadable segment, or a dynamically linkable information, or an interpretable information, or auxiliary information, etc.The rest of the information in the Program Header are flags of 1 byte or 2 bytes. I can be assumed that tampering this information will not be of interest for the malware developers.

3.12.3 Section information

There are many sections in an ELF file. Each section is for a specific type of information. The types of sections can be for a *Program data*, Symbol table, String table, Symbol Hash table, Dynamic Linking Information,

Dynamic Linker Symbol table, Array of Constructors, Array of Destructors, etc. For each section, there is information regarding whether the section is writable, or executable, might be merged, contained null-terminated strings, etc. Then, there is information regarding the various addresses for the different information. So, this is a section where various information can be altered by makers of malware.

3.12.4 ELF dataset

The most critical component of this project is the data on which the model is trained. The difficulty is to obtain appropriate data. The malicious ELF files are obtained from McAfee Research Unit in India through connections of the author with the director of the center. The good ELF files are collected from various systems. The final dataset contained files as provided in the Table 3.1.

So, the dataset contained 2315 benign ELF files and 3428 malicious ELF files. As we can see that there are reasonable number of files in each category. Also, the dataset is reasonably balanced. So, no special treatment is done for balancing the data. Once the data is obtained, it is required to create the training set and testing set. It is planned that 90% of the data would be used as the training set and 10% of the data would be used as the test set.

Table 3.1   Final dataset.

Benign		2,315
Malicious	Backdoor	744
	Botnet	782
	DDOS	799
	Trojan	508
	Virus	595
	TOTAL	3,428

Table 3.2   Distribution of files.

File type		Training set	Test set
Benign		2,078	237
Malicious	Backdoor	676	68
	Botnet	713	69
	DDOS	725	74
	Trojan	461	47
	Virus	539	56
	TOTAL	3,114	314

3.12.5 Distribution of dataset

The distribution of dataset used for training and testing the model is given in Table 3.2.

Listing 3.15 shows the function which is used to check whether the available files are valid ELF files.

```
1  #Python Function to test whether a File is a valid ELF File
2  from elftools.elf.elffile import ELFFile
3  def isELFFile(file):
4      ''' Function Name: isELFFile
5          Purpose: To check whether a file is an ELF File
6          Input Parameters:
7              1. file: File to be checked
8          Output Parameters:
9              1. Boolean: True if File is an ELF File, False
       otherwise
10         '''
11     with open(file, 'rb') as elffile:
12         returnValue = True
13
14         try:
15             ELFFile(elffile) # The constructor of the
       ELFFile class checks whether the magic number for an ELF
       File exists in the provided file
16
17         except:
18             returnValue = False
19
20         finally:
21             elffile.close()
22
23     return returnValue
```

Listing 3.15   ELF check functions.

The final distribution of valid ELF files is as given in Table 3.3.

3.13 Feature Engineering andMachine Learning Classification

It is first required to extract features of the ELF files. All the information in the File Header, Program Header, and Section Information are not numeric data. There are many textual data as well. As almost all the textual data is available in binary form, it is not possible to make any sort of transformation

Table 3.3   Final distribution of valid ELF files.

File type		Training set	Test set
Benign		2,038	222
Malicious	Backdoor	589	62
	Botnet	616	61
	DDOS	642	71
	Trojan	369	40
	Virus	452	48
	TOTAL	2,668	282

like one-hot encoding, label encoding, etc., on this textual data. So, decision is to only use the numeric data extracted for the purpose of building the model.

As can be seen from the code provided in the subsequent section, more than 2500 features are extracted from the available ELF files (i.e., for each file, there are more than 2500 distinct pieces of information). So, every ELF file in the training set and the test set and the subsequent files to be tested could be evaluated on more than 2500 features available in the file (It needs noting that it is quite possible that all the files to be tested (whether in the test set or otherwise) may not contain all the features used in the training of the model). However, before a file is evaluated using the model, it is necessary to transform the file into a format so that all the features of the file are available as a vector.

The model needed is a Multi-Class Classification model. To build the model, many algorithms are tried including Multi-Level Logistic Regression, Support Vector Machine, Decision Trees. Finally, Random Forest algorithm is selected as the model built using Random Forest Algorithm produced the maximum accuracy.

To test the efficacy of the model, the confusion matrix is used. Confusion matrix gives a precise idea of the percentage of False-positives. We know that True-positives and false-Positives are the desirable results. The True-negatives are not desirable. However, they are less risky as if a file not containing a malware is classified as a file containing a malware, the system does not face any harm. The main concerns are the False-positives where a file containing a malware is classified as a benign file.

The model built using the Random Forest algorithm produces the least number of False-positives on both the training set and test set. Further, it is seen that the accuracy achieved on the training set is 99%. While the accuracy achieved on the test set is 87%. So, it concludes that the model has a certain degree of overfitting. However, it can be reasonably argued that the model does not suffer from a bias. To overcome the overfitting, the number of features used to build the model could be reduced using techniques like principal component analysis, etc. This has been kept as a future work on this project.

3.14 Building the Model

Now we discuss the code used to build the model.

3.14.1 Constants used

The following constants are used in the code.

```
1  #Constants used in the code
2  # Constants
3  EMPTY_SECTION_NAME_SUBSTITUTE = "S"
4  FILE_NAME_COLUMN_NAME = "FileName"
5  FILE_TYPE_COLUMN_NAME = 'FileType'
6  NUMERIC_COLUMN_IDENTIFIER = "N"
7  MALWARE_SECTION_NAME_PREFIX = "M"
8
9  BENIGN_FILE = 0
10 BACKDOOR_FILE = 1
11 BOTNET_FILE = 2
12 DDOS_FILE = 3
13 TROJAN_FILE = 4
14 VIRUS_FILE = 5
```

3.14.2 Functions used to extract information from ELF files

The following five functions are used to extract the information from the ELF files.

extractInformationFromADirectory: This function takes a single file name or a set of file names and returns all the information extracted from all the ELF files as a list. The following listing shows the corresponding code.

```
1  #Function extractInformationFromADirectory()
2
3  import glob
4
5  def extractInformationFromADirectory(inputDirectory):
6      # Declare Counters
7      numberOfValidFiles = 0
8      numberOfInvalidFiles = 0
```

```
 9    numberOfFilesWithNoFileHeader = 0
10    numberOfFilesWithNoProgramHeader = 0
11    numberOfFilesWithNoSections = 0
12
13    # Create an Empty List to hold all the features of all
      the files
14    returnValue = []
15
16    # Loop through all the files in the Input Directory
17    for file in glob.glob(inputDirectory):
18        # Create an empty Dictionary
19        oneFileFeatures = {}
20
21        try:
22            # Read the file and extract the features
23            fileFeatures, validFlag, hasFileHeader,
      hasProgramHeader, hasSections = ExtractFileDetails(file)
24            if(validFlag and len(fileFeatures.keys()) > 0):
25                numberOfValidFiles = numberOfValidFiles + 1
26
27                # Add the File Features to the main
      Dictionary
28                oneFileFeatures.update(fileFeatures)
29
30                # Add the entry to the return value
31                returnValue.append(oneFileFeatures)
32            else:
33                if hasFileHeader == False:
34                    numberOfFilesWithNoFileHeader =
      numberOfFilesWithNoFileHeader + 1
35                else:
36                    if hasProgramHeader == False:
37                        numberOfFilesWithNoProgramHeader =
      numberOfFilesWithNoProgramHeader + 1
38                    else:
39                        if hasSections == False:
40                            numberOfFilesWithNoSections =
      numberOfFilesWithNoSections + 1
41                        else:
42                            numberOfInvalidFiles =
      numberOfInvalidFiles + 1
43
44        except:
45            # Check if the file is a valid ELF File
46            if isELFFile(file):
47                pass
```

```
48        else:
49            numberOfInvalidFiles = numberOfInvalidFiles
      + 1
50
51    return (returnValue, numberOfValidFiles,
      numberOfInvalidFiles, numberOfFilesWithNoFileHeader,
      numberOfFilesWithNoProgramHeader,
      numberOfFilesWithNoSections)
```

ExtractFileDetails: This function takes a single ELF file and returns all the information of the ELF file. The following Listing 3.16 shows the corresponding function.

```
1  #Function ExtractFileDetails()
2  def ExtractFileDetails(file):
3      ''' Function Name: ExtractFileDetails
4          Purpose: To extract the details of all the
      information of an ELF File.
5                  The information in a ELF File is stored as
      File Header, Program Header and Section Information.
6                  There is one set of information for File
      Header and Program Header.
7                  However, there can be many sections in an
      ELF File.
8          Input Parameters:
9              1. file: A File from which the information has
      to be extracted.
10                 The file may be a valid ELF File or not
      a valid ELF File.
11         Output Parameters:
12             1. A Dictionary containing the attributes as the
      key and their values as the value.
13                 In case, no information can be extracted from
      the ELF File or the file is not a valid ELF File, an
      empty dictionary is returned.
14             2. Boolean value indication whether the provides
      file was a valid ELF File or not.
15                 True is the provide was an ELF File, False
      otherwise.
16     '''
17     returnValue = {}
18     validELFFile = True
19     hasFileHeader = True
20     hasProgramHeader = True
21     hasSections = True
22
```

```
23    try :
24        with open ( file , 'rb' ) as elffile :
25            try :
26                eFile = ELFFile ( elffile )
27
28                # Extract the File Header Information
29                fileHeader = ExtractFileHeader ( eFile )
30                if ( len ( fileHeader ) > 0 ) :
31                    returnValue . update ( fileHeader ) # Add the
        attributes to dictionary
32                else :
33                    hasFileHeader = False
34                    raise Exception ()
35
36                # Extract the Program Header Information
37                segmentDetails = ExtractSegmentDetails ( eFile
        )
38                if ( len ( segmentDetails ) > 0 ) :
39                    returnValue . update ( segmentDetails ) # Add
        the attributes to dictionary
40                else :
41                    hasProgramHeader = False
42                    raise Exception ()
43
44                sectionDetails = ExtractSectionDetails ( eFile
        )
45                if ( len ( sectionDetails ) > 0 ) :
46                    returnValue . update ( sectionDetails ) # Add
        the attributes to dictionary
47                else :
48                    hasSections = False
49                    raise Exception ()
50
51            except :
52                validELFFile = False
53
54            finally :
55                elffile . close ()
56    except :
57        pass
58
59    return ( returnValue , validELFFile , hasFileHeader ,
    hasProgramHeader , hasSections )
```

Listing 3.16 Extract File details.

ExtractFileHeader: This function takes a single ELF file and returns all the information from the File Header of the ELF file. The following Listing 3.17 shows the corresponding function.

```
1  #Function ExtractFileHeader()
2  def ExtractFileHeader(elffile):
3      ''' Function Name: ExtractFileHeader
4          Purpose: To extract the File Header of the ELF File.
5          Input Parameters:
6              1. elffile: A valid ELF File from which the
        Segment information has to be extracted
7          Output Parameters:
8              1. A Dictionary containing the attributes as the
        key and their values as the value
9                  In case, the ELF File has no File Header, an
        empty dictionary is returned (This is not possible for
        valid ELF Files)
10         '''
11     returnValue = {} # Dictionary to hold the unique
        attributes the header
12
13     # The header information can be obtained as a dictionary
        .
14     # However, there are some dictionaries inside this
        dictionary.
15     # So, we create a flat structure taking out all the
        unique attributes and form a dictionary.
16     for key, value in elffile.header.items():
17         # If the value for the key is a dictionary, then
        loop through all the attributes of this dictionary to
        collect the features.
18         # The approach is simplistic as I do not go for a
        recurssive function as it is known that there can be
        only one aditional level of dictionary.
19         # TODO: There is some hardcoding here. Will be
        removed later.
20         if key == "e_ident":
21             for insideKey, insideValue in value.items():
22                 if insideKey == "EI_MAG":
23                     pass
24                 else:
25                     returnValue[insideKey] = insideValue
26         else:
27             returnValue[key] = value
28
```

```
29        return returnValue
```

Listing 3.17   Extract File Header.

ExtractSegmentDetails: This function takes a single ELF file and returns all the information from the Segment Header of the ELF file. The following Listing 3.18 shows the corresponding function.

```
1  #Function ExtractSegmentDetails()
2  def ExtractSegmentDetails(elffile):
3      ''' Function Name: ExtractSegmentDetails
4          Purpose: To extract the details of all the segments
   in the ELF File.
5                      The segment information is the Program
   Header of a ELF File.
6          Input Parameters:
7              1. elffile: A valid ELF File from which the
   Segment information has to be extracted
8          Output Parameters:
9              1. A Dictionary containing the attributes as the
   key and their values as the value
10                  In case, the ELF File has no segments, an
   empty dictionary is returned (This is not possible for
   valid ELF Files)
11      '''
12      returnValue = {} # Dictionary to hold the unique
   attributes of all the segments
13
14      # Check if any segment information exists in the file
15      # If it does, collect all the information of all the
   attributes in all the segments
16      if(elffile.num_segments() > 0):
17          prefixDict = {} # Dictionary to hold the unique
   segment attributes for all the segment names
18
19          # Each ELF File may have one or more number of
   segments
20          # Loop through all the segments
21          for segment in elffile.iter_segments():
22              prefix = ""
23
24              # In each segment, there can be one or more
   number of attributes
25              # The attribute "p_type" contains the segment
   name
26              # Under each segment, the attributes may be
   different from the other segments
```

```
27        for attribute in segment.header:
28            # The segment name is stored in the
     attribute "p_type"
29            # So, we prefix the segment name to all the
     other attributes to uniquely identify each attribute for
     all the segments
30            if(attribute == 'p_type'):
31                ctr = 0
32                prefix = segment.header[attribute]
33                # Here we check if 2 or more segments
     have the same name
34                # If there are 2 or more segments with
     the same name,
35                # then each of the segments are uniquely
     identified by addng a running counter to the end of the
     segment name
36                while True:
37                    if prefix in prefixDict:
38                        ctr = ctr + 1
39                        prefix = (segment.header[
     attribute] + "-" + str(ctr))
40                    else:
41                        break
42
43                prefixDict[prefix] = 1 # Keep a note of
     the segments processed so far
44            else:
45                # Create a key as the "<segment name>-<
     attribute name>"
46                # And add it to the unique list of
     attributes
47                key = prefix + "-" + attribute
48                returnValue[key] = segment.header[
     attribute]
49
50    return returnValue
```

Listing 3.18   Extract Segment details.

ExtractSectionDetails: This function takes a single ELF file and returns all the information from all the Sections of the ELF file. The following Listing 3.19 shows the corresponding function.

```
1 #def ExtractSectionDetails(elffile):
2     ''' Function Name: ExtractSegmentDetails
3         Purpose: To extract the details of all the segments
     in the ELF File.
```

```
4                          The segment information is the Program
      Header of a ELF File.
5           Input Parameters:
6                 1. elffile: A valid ELF File from which the
      Segment information has to be extracted
7           Output Parameters:
8                 1. A Dictionary containing the attributes as the
      key and their values as the value
9                    In case, the ELF File has no segments, an
      empty dictionary is returned (This is not possible for
      valid ELF Files)
10        '''
11    returnValue = {}
12
13    # ELF Files with malware contain Section Names which
      have been tampered.
14    # If such a Section Name is found, then we store the
      tampered Section Name as a feature.
15    # To be able to store these in unique attributes, we set
      a counter across the file.
16    malwareSectionNameCounter = 0
17
18    # Check if any section information exists in the file
19    # If it does, collect all the information of all the
      attributes in all the sections
20    if(elffile.num_sections() > 0):
21        # Iterate through all the sections and gather the
      attributes.
22        for section in elffile.iter_sections():
23            # Every section has a name.
24            # Section Name has to be a valid ASCII string.
25            # Is the Section Name contains non-ASCII
      characters, then the file has been tampered.
26            if all((ord(char) > 32 and ord(char) < 128) for
      char in section.name):
27                sectionName = section.name
28            else:
29                # In case the Section Name contains non-
      ASCII characters, we store the Section Name as a
      features in out data set
30                malwareSectionNameCounter =
      malwareSectionNameCounter + 1
31                sectionName = MALWARE_SECTION_NAME_PREFIX +
      str(malwareSectionNameCounter)
32                returnValue[sectionName] = section.name
33
```

```
34          # Attributes of all the sections may have the
       same name.
35          # So, section name will be prefixed to the
       attribute name to form the key for the dictionary.
36          sectionName = sectionName.lstrip('.') # Remove
       leading dot ('.') from the section name
37          sectionName = sectionName.lstrip('_') # Remove
       leading underscores ('_') from the section name
38          sectionName = sectionName.replace('.', '-') #
       Remove all the dots ('.') and replace with a dash ('-')
39          sectionName = sectionName.strip()
40
41          if len(sectionName) == 0:
42              sectionName = EMPTY_SECTION_NAME_SUBSTITUTE
43
44          # Every section has a header.
45          # Iterate through all the attributes in the
       header of the section.
46          # The attributes of the section header are
       prefixed by the section name to form the attribute name.
47          for key, value in section.header.items():
48              attributeName = sectionName + "-" + key
49              returnValue[attributeName] = value
50
51      return returnValue
```

Listing 3.19   Extract Section details.

Using these functions, the features from the benign ELF files and the ELF files containing malware are extracted as shown in Listing 3.20.

```
1 #Code for Extracting the features from the ELF Files
2 benignFileFeatures, nValidFiles, nInvalidFiles,
     nFilesWithNoFileHeader, nFilesWithNoProgramHeader,
     nFilesWithNoSections = \
3 extractInformationFromADirectory("Data/benign_ELF/*")
4
5 print("BENIGN FILES:\nValid - %d\nInvalid - %d\nNo File
     Header - %d\nNo Program Header - %d\nNo Sections - %d" %
     \
6     (nValidFiles, nInvalidFiles, nFilesWithNoFileHeader,
     nFilesWithNoProgramHeader, nFilesWithNoSections))
7
8 BENIGN FILES:
9 Valid - 2038
10 Invalid - 30
11 No File Header - 0
```

```
12 No Program Header - 0
13 No Sections - 10
14
15 backdoorFileFeatures, nValidFiles, nInvalidFiles,
      nFilesWithNoFileHeader, nFilesWithNoProgramHeader,
      nFilesWithNoSections = \
16 extractInformationFromADirectory("Data/malware_ELF/Backdoor
      /*")
17
18 print("BACKDOOR FILES:\nValid - %d\nInvalid - %d\nNo File
      Header - %d\nNo Program Header - %d\nNo Sections - %d" %
      \
19       (nValidFiles, nInvalidFiles, nFilesWithNoFileHeader,
      nFilesWithNoProgramHeader, nFilesWithNoSections))
20
21 BACKDOOR FILES:
22 Valid - 589
23 Invalid - 36
24 No File Header - 0
25 No Program Header - 1
26 No Sections - 50
27
28 botnetFileFeatures, nValidFiles, nInvalidFiles,
      nFilesWithNoFileHeader, nFilesWithNoProgramHeader,
      nFilesWithNoSections = \
29 extractInformationFromADirectory("Data/malware_ELF/Botnet/*"
      )
30
31 print("BOTNET FILES:\nValid - %d\nInvalid - %d\nNo File
      Header - %d\nNo Program Header - %d\nNo Sections - %d" %
      \
32       (nValidFiles, nInvalidFiles, nFilesWithNoFileHeader,
      nFilesWithNoProgramHeader, nFilesWithNoSections))
33
34 BOTNET FILES:
35 Valid - 616
36 Invalid - 64
37 No File Header - 0
38 No Program Header - 0
39 No Sections - 33
40
41 ddosFileFeatures, nValidFiles, nInvalidFiles,
      nFilesWithNoFileHeader, nFilesWithNoProgramHeader,
      nFilesWithNoSections = \
42 extractInformationFromADirectory("Data/malware_ELF/Ddos/*")
43
```

```
44  print("DDOS FILES:\nValid - %d\nInvalid - %d\nNo File Header
        - %d\nNo Program Header - %d\nNo Sections - %d" % \
45      (nValidFiles, nInvalidFiles, nFilesWithNoFileHeader,
        nFilesWithNoProgramHeader, nFilesWithNoSections))
46
47  DDOS FILES:
48  Valid - 642
49  Invalid - 74
50  No File Header - 0
51  No Program Header - 0
52  No Sections - 9
53
54  trojanFileFeatures, nValidFiles, nInvalidFiles,
        nFilesWithNoFileHeader, nFilesWithNoProgramHeader,
        nFilesWithNoSections = \
55  extractInformationFromADirectory("Data/malware_ELF/Trojan/*"
        )
56
57  print("TROJAN FILES:\nValid - %d\nInvalid - %d\nNo File
        Header - %d\nNo Program Header - %d\nNo Sections - %d" %
        \
58      (nValidFiles, nInvalidFiles, nFilesWithNoFileHeader,
        nFilesWithNoProgramHeader, nFilesWithNoSections))
59
60  TROJAN FILES:
61  Valid - 369
62  Invalid - 46
63  No File Header - 0
64  No Program Header - 0
65  No Sections - 46
66
67  virusFileFeatures, nValidFiles, nInvalidFiles,
        nFilesWithNoFileHeader, nFilesWithNoProgramHeader,
        nFilesWithNoSections = \
68  extractInformationFromADirectory("Data/malware_ELF/Virus/*")
69
70  print("VIRUS FILES:\nValid - %d\nInvalid - %d\nNo File
        Header - %d\nNo Program Header - %d\nNo Sections - %d" %
        \
71      (nValidFiles, nInvalidFiles, nFilesWithNoFileHeader,
        nFilesWithNoProgramHeader, nFilesWithNoSections))
72
73  VIRUS FILES:
74  Valid - 452
75  Invalid - 29
76  No File Header - 0
```

```
77 No Program  Header  -  0
78 No Sections  -  58
```

Listing 3.20   Extract Features from all files.

3.15 Extract the Unique List of Keys for All the Files

From the last steps, we have six dictionaries containing all the attributes extracted from all the clean ELF files and from all the ELF files containing malware. In this step, a list of all the unique attributes is prepared. Listing 3.21 shows the code for creating the set of unique features.

```
1  #Code for creating the set of unique features across all
       files in the training data set
2  featureList = set()
3
4  for i in benignFileFeatures:
5      for k in i.keys():
6          if type(k) == int:
7              featureList.add(NUMERIC_COLUMN_IDENTIFIER + str(
   k))
8          else:
9              featureList.add(k)
10
11 for i in backdoorFileFeatures:
12     for k in i.keys():
13         if type(k) == int:
14             featureList.add(NUMERIC_COLUMN_IDENTIFIER + str(
   k))
15         else:
16             featureList.add(k)
17
18 for i in botnetFileFeatures:
19     for k in i.keys():
20         if type(k) == int:
21             featureList.add(NUMERIC_COLUMN_IDENTIFIER + str(
   k))
22         else:
23             featureList.add(k)
24
25 for i in ddosFileFeatures:
26     for k in i.keys():
27         if type(k) == int:
28             featureList.add(NUMERIC_COLUMN_IDENTIFIER + str(
   k))
```

```
29        else:
30            featureList.add(k)
31
32 for i in trojanFileFeatures:
33     for k in i.keys():
34         if type(k) == int:
35             featureList.add(NUMERIC_COLUMN_IDENTIFIER + str(
    k))
36         else:
37             featureList.add(k)
38
39 for i in virusFileFeatures:
40     for k in i.keys():
41         if type(k) == int:
42             featureList.add(NUMERIC_COLUMN_IDENTIFIER + str(
    k))
43         else:
44             featureList.add(k)
```

Listing 3.21 Code for creating the set of unique features.

3.16 Create a Data Frame

Now we create a data frame containing all the keys and values for all the files in the training dataset. To be able to do this, we use a function as shown in Listing 3.22.

```
1 #Function fillDataInDataFrame()
2 import pandas as pd
3
4 def fillDataInDataFrame(featureDictionary, p_featureList,
    fileType):
5     # Create an Empty DataFrame object
6     df = pd.DataFrame()
7
8     for record in featureDictionary:
9         # Create an empty Dictionary
10        oneRecord = {}
11
12        # Initialise all the columns
13        for colName in p_featureList:
14            oneRecord[colName] = pd.to_numeric(0, downcast=
    'integer')
15
16        # Loop through all the features in a record
```

```
17          for  k  in  record.keys():
18                  # Extract the column name from the record
19                  if type(k) == int:
20                      extractedColumnName =
     NUMERIC_COLUMN_IDENTIFIER + str(k)
21                  else:
22                      extractedColumnName = k
23
24              # Check if the column name exists in the feature
         list
25                  if extractedColumnName in p_featureList:
26                      # Extract the value for the key and store in
         the Dictionary
27                          if ((type(record[k]) == int) | (type(record[
     k]) == float)):
28                              oneRecord[extractedColumnName] = pd.
     to_numeric(record[k], downcast='integer')
29                          else:
30                              oneRecord[extractedColumnName] = record[
     k]
31
32          # Add column to mark Dependent Column as per the
         provided File Type
33          oneRecord[FILE_TYPE_COLUMN_NAME] = pd.to_numeric(
     fileType, downcast='integer')
34
35          # Add the Record to the Data Frame
36          df = df.append(oneRecord, ignore_index=True)
37
38      return df
```

Listing 3.22   Data frame generation function.

Now we use this function to prepare the data frame with labeled data the corresponding code is shown in Listing 3.23.

```
1  #Code to prepare the data frame of classified features
2  benignFileDF = fillDataInDataFrame(benignFileFeatures,
       featureList, BENIGN_FILE)
3
4  backdoorFileDF = fillDataInDataFrame(backdoorFileFeatures,
       featureList, BACKDOOR_FILE)
5
6  botnetFileDF = fillDataInDataFrame(botnetFileFeatures,
       featureList, BOTNET_FILE)
7
```

```
8  ddosFileDF = fillDataInDataFrame(ddosFileFeatures,
       featureList, DDOS_FILE)
9
10 trojanFileDF = fillDataInDataFrame(trojanFileFeatures,
       featureList, TROJAN_FILE)
11
12 virusFileDF = fillDataInDataFrame(virusFileFeatures,
       featureList, VIRUS_FILE)
13
14 # Form a single data frame of all the labelled data
15 df = pd.concat([benignFileDF, backdoorFileDF, botnetFileDF,
       ddosFileDF, trojanFileDF, virusFileDF], ignore_index=
       True)
16
17 df[FILE_TYPE_COLUMN_NAME] = pd.to_numeric(df[
       FILE_TYPE_COLUMN_NAME], downcast='integer')
```

Listing 3.23   Data frame preparation.

Now the data frame is ready. However, as we had done for JPEG File features, we remove any presence of CHR(0) so that reading the data is not truncated by the presence of CHR(0).

```
1 #Code to remove CHR(0)
2 # Remove NULL character from all the String
3 for col in df.columns:
4     if df[col].dtypes == 'object':
5         df[col].replace(chr(0), '', inplace = True)
6
7 df[FILE_TYPE_COLUMN_NAME].value_counts()
8
9  0    2038
10 3     642
11 2     616
12 1     589
13 5     452
14 4     369
15 Name: FileType, dtype: int64
```

As a last step for data preparation, we only retain the columns in the above formed data frame where the data type is numeric. Then, we can create the data frame of independent variables and the data frame of the dependent variable.

After we have done that, we save the list of columns. This is because we will need extract information for only these columns (or features) from the test dataset and/or from the data provided during production run.

```
1 #Code to retain the numeric columns
2 dfClean = df.select_dtypes([np.number])
3
4 X = dfClean.copy()
5 X = X.drop([FILE_TYPE_COLUMN_NAME], axis = 1)
6 y = dfClean[FILE_TYPE_COLUMN_NAME]
7
8 import pickle
9 pickle.dump(X, open("./XColumns", 'wb'))
```

3.17 Random Forest Model Generation

Creating the model using Random Forest algorithm for ELF file classification is like creating the model using the Random Forest algorithm for JPEG files. Listing 3.24 shows the code for Random Forest model generation.

```
1 # Code to form the classification model using Random Forest
       Algorithm
2 from sklearn.model_selection import cross_val_score
3 from sklearn.model_selection import RepeatedStratifiedKFold
4 from sklearn.ensemble import RandomForestClassifier
5
6 # Instantiate the Decision Tree Model
7 modelRF = RandomForestClassifier()
8
9 # Create the Random Forest Model
10 modelRF.fit(X, y)
11
12 # Save the Random Forest Model to a file
13 import pickle
14
15 pickle.dump(modelRF, open("./RFModelELFMalwareDetection",
       'wb'))
```

Listing 3.24   Classification model using Random Forest.

We test the results for the model on the training dataset as shown in Listing 3.25.

```
1 #Code to make predictions on the training data set using the
       model created using the Random Forest Algorithm.
2
3 from sklearn import metrics
4 import matplotlib.pyplot as plt
5 import seaborn as sns
```

```
 6
 7  # Make the predictions
 8  y_pred = modelRF.predict(X)
 9
10  # Generate the Confusion Matrix
11  cm = metrics.confusion_matrix(y, y_pred)
12
13  # Plot the Confusion Matrix
14  ax = plt.subplot()
15  sns.heatmap(cm, annot=True, fmt='g', ax=ax);
16
17  ax.set_xlabel('Predicted labels');
18  ax.set_ylabel('True labels');
19  ax.set_title('Confusion Matrix');
20  ax.xaxis.set_ticklabels(['Benign', 'Backdoor', 'Botnet',
        'DDOS', 'Trojan', 'Virus']);
21  ax.yaxis.set_ticklabels(['Benign', 'Backdoor', 'Botnet',
        'DDOS', 'Trojan', 'Virus']);
22
23  print("\n\nConfusion Classification Report\n")
24  print(metrics.classification_report(y, y_pred))
25
26  Confusion Classification Report
27
28                 precision    recall  f1-score   support
29
30            0       1.00      1.00      1.00      2038
31            1       0.96      0.99      0.98       589
32            2       1.00      0.97      0.99       616
33            3       0.99      1.00      0.99       642
34            4       0.93      0.96      0.94       369
35            5       0.98      0.94      0.96       452
36
37     accuracy                           0.99      4706
38    macro avg       0.98      0.98      0.98      4706
39 weighted avg       0.99      0.99      0.99      4706
```

Listing 3.25   Model results using training dataset.

Now, we test the model on the test dataset. Before we can make predictions on the test dataset, we must make the data from the test dataset suitable for being used by the model. Listing 3.26 shows the necessary code to prepare the test dataset before predictions can be made for the same.

```
 1  #Code to prepare the test data set before predictions can be
        made on them
```

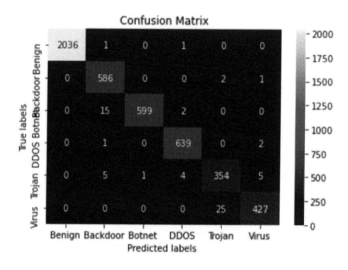

Figure 3.7   Confusion matrix of predictions on training data of ELF files using model developed using Random Forest algorithm.

```
2 benignTestFileFeatures, nValidFiles, nInvalidFiles,
    nFilesWithNoFileHeader, nFilesWithNoProgramHeader,
    nFilesWithNoSections = \
3 extractInformationFromADirectory("Data/TestData/benign_ELF/*
    ")
4
5 print("BENIGN FILES:\nValid - %d\nInvalid - %d\nNo File
    Header - %d\nNo Program Header - %d\nNo Sections - %d" %
    \
6      (nValidFiles, nInvalidFiles, nFilesWithNoFileHeader,
    nFilesWithNoProgramHeader, nFilesWithNoSections))
7
8 BENIGN FILES:
9 Valid - 222
10 Invalid - 10
11 No File Header - 0
12 No Program Header - 0
13 No Sections - 5
14
15 backdoorTestFileFeatures, nValidFiles, nInvalidFiles,
    nFilesWithNoFileHeader, nFilesWithNoProgramHeader,
    nFilesWithNoSections = \
```

```
16  extractInformationFromADirectory("Data/TestData/malware_ELF/
        Backdoor/*")
17
18  print("BACKDOOR FILES:\nValid - %d\nInvalid - %d\nNo File
        Header - %d\nNo Program Header - %d\nNo Sections - %d" %
        \
19          (nValidFiles, nInvalidFiles, nFilesWithNoFileHeader,
        nFilesWithNoProgramHeader, nFilesWithNoSections))
20
21  BACKDOOR FILES:
22  Valid - 62
23  Invalid - 2
24  No File Header - 0
25  No Program Header - 1
26  No Sections - 4
27
28  botnetTestFileFeatures, nValidFiles, nInvalidFiles,
        nFilesWithNoFileHeader, nFilesWithNoProgramHeader,
        nFilesWithNoSections = \
29  extractInformationFromADirectory("Data/TestData/malware_ELF/
        Botnet/*")
30
31  print("BOTNET FILES:\nValid - %d\nInvalid - %d\nNo File
        Header - %d\nNo Program Header - %d\nNo Sections - %d" %
        \
32          (nValidFiles, nInvalidFiles, nFilesWithNoFileHeader,
        nFilesWithNoProgramHeader, nFilesWithNoSections))
33
34  BOTNET FILES:
35  Valid - 61
36  Invalid - 6
37  No File Header - 0
38  No Program Header - 0
39  No Sections - 2
40
41  ddosTestFileFeatures, nValidFiles, nInvalidFiles,
        nFilesWithNoFileHeader, nFilesWithNoProgramHeader,
        nFilesWithNoSections = \
42  extractInformationFromADirectory("Data/TestData/malware_ELF/
        Ddos/*")
43
44  print("DDOS FILES:\nValid - %d\nInvalid - %d\nNo File Header
        - %d\nNo Program Header - %d\nNo Sections - %d" % \
45          (nValidFiles, nInvalidFiles, nFilesWithNoFileHeader,
        nFilesWithNoProgramHeader, nFilesWithNoSections))
46
```

```
47  DDOS FILES:
48  Valid - 71
49  Invalid - 3
50  No File Header - 0
51  No Program Header - 0
52  No Sections - 0
53
54  trojanTestFileFeatures, nValidFiles, nInvalidFiles,
        nFilesWithNoFileHeader, nFilesWithNoProgramHeader,
        nFilesWithNoSections = \
55  extractInformationFromADirectory("Data/TestData/malware_ELF/
        Trojan/*")
56
57  print("TROJAN FILES:\nValid - %d\nInvalid - %d\nNo File
        Header - %d\nNo Program Header - %d\nNo Sections - %d" %
        \
58        (nValidFiles, nInvalidFiles, nFilesWithNoFileHeader,
        nFilesWithNoProgramHeader, nFilesWithNoSections))
59
60  TROJAN FILES:
61  Valid - 40
62  Invalid - 3
63  No File Header - 0
64  No Program Header - 0
65  No Sections - 4
66
67  virusTestFileFeatures, nValidFiles, nInvalidFiles,
        nFilesWithNoFileHeader, nFilesWithNoProgramHeader,
        nFilesWithNoSections = \
68  extractInformationFromADirectory("Data/TestData/malware_ELF/
        Virus/*")
69
70  print("VIRUS FILES:\nValid - %d\nInvalid - %d\nNo File
        Header - %d\nNo Program Header - %d\nNo Sections - %d" %
        \
71        (nValidFiles, nInvalidFiles, nFilesWithNoFileHeader,
        nFilesWithNoProgramHeader, nFilesWithNoSections))
72
73  VIRUS FILES:
74  Valid - 48
75  Invalid - 4
76  No File Header - 0
77  No Program Header - 0
78  No Sections - 4
79
80  # Form the data frame
```

```
81  columnsInModel = X.columns
82
83  benignTestFileDF = fillDataInDataFrame(
        benignTestFileFeatures, set(columnsInModel), BENIGN_FILE
        )
84
85  backdoorTestFileDF = fillDataInDataFrame(
        backdoorTestFileFeatures, set(columnsInModel),
        BACKDOOR_FILE)
86
87  botnetTestFileDF = fillDataInDataFrame(
        botnetTestFileFeatures, set(columnsInModel), BOTNET_FILE
        )
88
89  ddosTestFileDF = fillDataInDataFrame(ddosTestFileFeatures,
        set(columnsInModel), DDOS_FILE)
90
91  trojanTestFileDF = fillDataInDataFrame(
        trojanTestFileFeatures, set(columnsInModel), TROJAN_FILE
        )
92
93  virusTestFileDF = fillDataInDataFrame(virusTestFileFeatures,
        set(columnsInModel), VIRUS_FILE)
94
95  # Combine the data frames formed above
96  dfTest = pd.concat([benignTestFileDF, backdoorTestFileDF,
        botnetTestFileDF, ddosTestFileDF, trojanTestFileDF,
        virusTestFileDF], ignore_index=True)
97
98  testDF[FILE_TYPE_COLUMN_NAME] = pd.to_numeric(testDF[
        FILE_TYPE_COLUMN_NAME], downcast='integer')
99
100 # Clean the data
101 for col in dfTest.columns:
102     if dfTest[col].dtypes == 'object':
103         dfTest[col].replace(chr(0), '', inplace = True)
104
105 # Create the X and y data frames
106 X_test = dfTest.copy()
107 X_test = X_test.drop([FILE_TYPE_COLUMN_NAME], axis = 1)
108 Y_test = dfTest[FILE_TYPE_COLUMN_NAME]
```

Listing 3.26   Code to prepare test data.

Listing 3.27 states the mechanism to make predictions on the test dataset using the model created using Random Forest algorithm and check the results.

We see that we get an accuracy of about 89% on the test dataset. Figures 3.7 and 3.8 show the corresponding confusion matrices generated using training and testing data, respectively with Random Forest classifier for ELF files.

```
1  #Code to make predictions on the test data and evaluate the
      results
2  # Make the predictions
3  y_pred_test = modelRF.predict(X_test)
4
5  # Generate the Confusion Matrix
6  cm = metrics.confusion_matrix(y_test, y_pred_test)
7
8  # Plot the Confusion Matrix
9  ax = plt.subplot()
10 sns.heatmap(cm, annot=True, fmt='g', ax=ax);
11
12 ax.set_xlabel('Predicted labels');
13 ax.set_ylabel('True labels');
14 ax.set_title('Confusion Matrix');
15 ax.xaxis.set_ticklabels(['Benign', 'Backdoor', 'Botnet',
      'DDOS', 'Trojan', 'Virus']);
16 ax.yaxis.set_ticklabels(['Benign', 'Backdoor', 'Botnet',
      'DDOS', 'Trojan', 'Virus']);
```

Listing 3.27   Model results using testing dataset.

3.18 Outcomes from the Model

We see that both the models developed demonstrate a dissimilar accuracy in training and during testing. We notice that we get about 99% accuracy on the training dataset. The accuracy drops of about 89% on the test dataset. So, we can conclude that the model suffers from overfitting. The overfitting can be attributed to the complexity of the model. So, the model has a high variance. We can solve the overfitting problem by conducting a principal component analysis (PCA) on the extracted features. This will help in simplifying the model and thus the variance will reduce.

3.19 Creating the Web Service

Now that we have our model, we can create the web service through which the model can be accessed by any client. Creating a web service will allow any kind of client application to use the model.

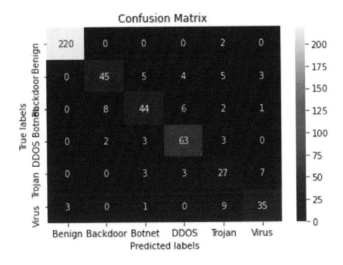

Figure 3.8 Confusion matrix of predictions on test data of ELF files using model developed using Random Forest algorithm.

Listing 3.28 shows the code for the web service. The web service takes a file as an input. The web service evaluates whether the file is a valid ELF file. For every valid ELF file, the web service reports whether the file contains virus or is a clean ELF file. In case the ELF file is found to have a malware, the web service reports the kind of malware available in the ELF file.

```
Code for the Web Service

#!/usr/bin/env python
# -*- coding: UTF-8 -*-

import pandas as pd
import os
import cherrypy
import pickle
# For ELF File Feature Extraction
from elftools.elf.elffile import ELFFile
from elftools.elf.segments import Segment

config = {
    'global' : {
        'server.socket_host' : '127.0.0.1',
```

```
18          'server.socket_port' : 8080,
19          'server.thread_pool' : 8,
20          'server.max_request_body_size' : 0,
21          'server.socket_timeout' : 60
22      }
23  }
24
25  class App:
26
27      def __init__(self):
28          self.RFmodel = pickle.load(open
29  ('RFModelELFMalwareDetection', 'rb'))
30          self.featureList = pickle.load(open('FeatureList',
        'rb'))
31          self.EMPTY_SECTION_NAME_SUBSTITUTE = "S"
32          self.FILE_NAME_COLUMN_NAME = "FileName"
33          self.FILE_TYPE_COLUMN_NAME = 'FileType'
34          self.NUMERIC_COLUMN_IDENTIFIER = "N"
35          self.MALWARE_SECTION_NAME_PREFIX = "M"
36
37      def ExtractFileHeader(self, elffile):
38          returnValue = {} # Dictionary to hold the unique
        attributes the header
39
40          # The header information can be obtained as a
        dictionary.
41          # However, there are some dictionaries inside this
        dictionary.
42          # So, we create a flat structure taking out all the
        unique attributes and form a dictionary.
43          for key, value in elffile.header.items():
44              # If the value for the key is a dictionary, then
        loop through all the attributes of this dictionary to
        collect the features.
45              # The approach is simplistic as I do not go for
        a recurssive function as it is known that there can be
        only one aditional level of dictionary.
46              # TODO: There is some hardcoding here. Will be
        removed later.
47              if key == "e_ident":
48                  for insideKey, insideValue in value.items():
49                      if insideKey == "EI_MAG":
50                          pass
51                      else:
52                          returnValue[insideKey] = insideValue
53              else:
```

```
54              returnValue[key] = value
55
56          return returnValue
57
58      def ExtractSegmentDetails(self, elffile):
59          returnValue = {} # Dictionary to hold the unique
            attributes of all the segments
60
61              # Check if any segment information exists in the
            file
62              # If it does, collect all the information of all the
            attributes in all the segments
63          if(elffile.num_segments() > 0):
64              prefixDict = {} # Dictionary to hold the unique
            segment attributes for all the segment names
65
66              # Each ELF File may have one or more number of
            segments
67              # Loop through all the segments
68          for segment in elffile.iter_segments():
69              prefix = ""
70
71                  # In each segment, there can be one or more
            number of attributes
72                  # The attribute "p_type" contains the
            segment name
73                  # Under each segment, the attributes may be
            different from the other segments
74              for attribute in segment.header:
75                  # The segment name is stored in the
            attribute "p_type"
76                  # So, we prefix the segment name to all
            the other attributes to uniquely identify each attribute
            for all the segments
77              if(attribute == 'p_type'):
78                  ctr = 0
79                  prefix = segment.header[attribute]
80                  # Here we check if 2 or more
            segments have the same name
81                  # If there are 2 or more segments
            with the same name,
82                  # then each of the segments are
            uniquely identified by addng a running counter to the
            end of the segment name
83              while True:
84                  if prefix in prefixDict:
```

```
85                                    ctr = ctr + 1
86                                    prefix = (segment.header[
      attribute] + "-" + str(ctr))
87                                else:
88                                    break
89
90                                prefixDict[prefix] = 1 # Keep a note
      of the segments processed so far
91                            else:
92                                # Create a key as the "<segment name
      >-<attribute name>"
93                                # And add it to the unique list of
      attributes
94                                key = prefix + "-" + attribute
95                                returnValue[key] = segment.header[
      attribute]
96
97          return returnValue
98
99      def ExtractSectionDetails(self, elffile):
100         returnValue = {}
101
102         # ELF Files with malware contain Section Names which
      have been tampered.
103         # If such a Section Name is found, then we store the
      tampered Section Name as a feature.
104         # To be able to store these in unique attributes, we
      set a counter across the file.
105         malwareSectionNameCounter = 0
106
107         # Check if any section information exists in the
      file
108         # If it does, collect all the information of all the
      attributes in all the sections
109         if(elffile.num_sections() > 0):
110             # Iterate through all the sections and gather
      the attributes.
111             for section in elffile.iter_sections():
112                 # Every section has a name.
113                 # Section Name has to be a valid ASCII
      string.
114                 # Is the Section Name contains non-ASCII
      characters, then the file has been tampered.
115                 if all((ord(char) > 32 and ord(char) < 128)
      for char in section.name):
116                     sectionName = section.name
```

```
117            else:
118                        # In case the Section Name contains non-
       ASCII characters, we store the Section Name as a
       features in out data set
119                        malwareSectionNameCounter =
       malwareSectionNameCounter + 1
120                        sectionName = self.
       MALWARE_SECTION_NAME_PREFIX + str(
       malwareSectionNameCounter)
121                        returnValue[sectionName] = section.name
122
123                    # Attributes of all the sections may have
       the same name.
124                    # So, section name will be prefixed to the
       attribute name to form the key for the dictionary.
125                    sectionName = sectionName.lstrip('.') #
       Remove leading dot ('.') from the section name
126                    sectionName = sectionName.lstrip('_') #
       Remove leading underscores ('_') from the section name
127                    sectionName = sectionName.replace('.', '-')
       # Remove all the dots ('.') and replace with a dash
       ('-')
128                    sectionName = sectionName.strip()
129
130                    if len(sectionName) == 0:
131                        sectionName = self.
       EMPTY_SECTION_NAME_SUBSTITUTE
132
133                    # Every section has a header.
134                    # Iterate through all the attributes in the
       header of the section.
135                    # The attributes of the section header are
       prefixed by the section name to form the attribute name.
136                    for key, value in section.header.items():
137                        attributeName = sectionName + "-" + key
138                        returnValue[attributeName] = value
139
140         return returnValue
141
142     def ExtractFileDetails(self, file):
143         returnValue = {}
144         validELFFile = True
145         hasFileHeader = True
146         hasProgramHeader = True
147         hasSections = True
148
```

```
149        try:
150            with open(file, 'rb') as elffile:
151                try:
152                    eFile = ELFFile(elffile)
153
154                    # Extract the File Header Information
155                    fileHeader = self.ExtractFileHeader(
    eFile)
156                    if(len(fileHeader) > 0):
157                        returnValue.update(fileHeader) # Add
    the attributes to dictionary
158                    else:
159                        hasFileHeader = False
160                        raise Exception()
161
162                    # Extract the Program Header Information
163                    segmentDetails = self.
    ExtractSegmentDetails(eFile)
164                    if(len(segmentDetails) > 0):
165                        returnValue.update(segmentDetails) #
    Add the attributes to dictionary
166                    else:
167                        hasProgramHeader = False
168                        raise Exception()
169
170                    sectionDetails = self.
    ExtractSectionDetails(eFile)
171                    if(len(sectionDetails) > 0):
172                        returnValue.update(sectionDetails) #
    Add the attributes to dictionary
173                    else:
174                        hasSections = False
175                        raise Exception()
176
177                except:
178                    validELFFile = False
179
180                finally:
181                    elffile.close()
182        except:
183            pass
184
185        return (returnValue, validELFFile, hasFileHeader,
    hasProgramHeader, hasSections)
186
187    def isELFFile(self, file):
```

```
188        with open( file , 'rb') as elffile:
189            returnValue = True
190
191            try:
192                ELFFile( elffile ) # The constructor of the
      ELFFile class checks whether the magic number for an ELF
      File exists in the provided file
193
194            except:
195                returnValue = False
196
197            finally:
198                elffile . close ()
199
200        return returnValue
201
202
203    def fillDataInDataFrame( self , featureDictionary ):
204        # Create an Empty DataFrame object
205        df = pd . DataFrame ()
206
207        for record in featureDictionary:
208            # Create an empty Dictionary
209            oneRecord = {}
210
211            # Initialise all the columns
212            for colName in self . featureList:
213                oneRecord [ colName ] = pd . to_numeric (0 ,
      downcast='integer ')
214
215            # Loop through all the features in a record
216            for k in record . keys ():
217                # Extract the column name from the record
218                if type (k) == int:
219                    extractedColumnName = self .
      NUMERIC_COLUMN_IDENTIFIER + str (k)
220                else:
221                    extractedColumnName = k
222
223                # Check if the column name exists in the
      feature list
224                if extractedColumnName in self . featureList:
225                    # Extract the value for the key and
      store in the Dictionary
226                    if (( type ( record [k]) == int) | ( type (
      record [k]) == float )):
```

```
227                              oneRecord[extractedColumnName] = pd.
          to_numeric(record[k], downcast='integer')
228                          else:
229                              oneRecord[extractedColumnName] =
          record[k]
230
231              # Add the Record to the Data Frame
232              df = df.append(oneRecord, ignore_index=True)
233
234          return df
235
236      def prepareFileForAnalysis(self, ELFFileName):
237          returnValue = 0
238          X = None
239
240          testFeatures, validELFFile, hasFileHeader,
          hasProgramHeader, hasSections = self.ExtractFileDetails(
          ELFFileName)
241          if validELFFile == False:
242              returnValue = -1
243          elif hasFileHeader == False:
244              returnValue = -2
245          elif hasProgramHeader == False:
246              returnValue = -3
247          elif hasSections == False:
248              returnValue = -4
249          else:
250              testDF = self.fillDataInDataFrame(testFeatures)
251
252              # Remove NULL character from all the String
253              for col in testDF.columns:
254                  if testDF[col].dtypes == 'object':
255                      testDF[col].replace(chr(0), '', inplace
          = True)
256
257          return testDF, returnValue
258
259      @cherrypy.expose
260      def upload(self, ufile):
261          upload_path = os.path.normpath('./data/')
262          upload_file = os.path.join(upload_path, ufile.
          filename)
263          size = 0
264
265          returnValue = 0
266
```

```
267        with open(upload_file, 'b') as out:
268            while True:
269                data = ufile.file.read(8192)
270                if not data:
271                    break
272                out.write(data)
273                size += len(data)
274
275        # Check whether File is an Image File
276        if (self.isELFFile(upload_file) == False):
277            returnValue = -1 # Not an ELF File
278        else:
279            X, fileStatus = self.prepareFileForAnalysis(
       upload_file)
280            if fileStatus < 0:
281                returnValue = fileStatus
282            else:
283                returnValue = self.RFmodel.predict(X)[0]
284
285        out = '''
286            returnValue: {}
287            length: {}
288            filename: {}
289            mime-type: {}
290            '''.format(returnValue, size, ufile.
       filename, ufile.content_type, data)
291
292        return out
293
294
295 if __name__ == '__main__':
296    cherrypy.quickstart(App(), '/', config)
```

Listing 3.28   Web service for ELF.

3.20 Conclusion

We find that it is possible to rely on machine learning models for detecting malware. This work demonstrates the detection of malware in JPEG files and ELF files. A similar strategy can be used for other file types like Windows Executable files, PDF files, etc. One can improve the accuracy of the models discussed in this work if a given dataset is available for building the model.

The dataset used in this work can be stated as a small dataset for creating a commercial product for malware detection. One can further improve the models by creating models using neural networks.

3.21 Acknowledgments

We would like to thank Professor Sandeep K. Shukla for approving this project and encouraging us to study this subject.

4

New Age Attack Vectors – JPEG Images Machine Learning-based Solution for the Detection of Malicious JPEG Images

Shankar Kashamshetty and Kunal Chawla

E-mail: kshankar14@gmail.com; kunaalchawla@yahoo.com

Abstract

Cyberattacks against individuals, businesses, and organizations have increased in recent years. Cybercriminals are always looking for valuable vectors to distribute malware to victims to initiate an attack. Millions of people use images every day, and many users believe images are safe for use; still, various images can hold a malicious payload and execute unsafe actions. JPEG is the trendy image format, mainly due to its lossy compression. It is used by nearly everybody, from persons to large organizations, and almost every device (e.g., digital cameras, smartphones, social media, etc.). As of their safe reputation, enormous use, and high possibility for exploitation, JPEG images are used by cybercriminals as an attack vector. Although machine learning methods are useful at identifying known and unknown malware in different domains, machine learning techniques have not been used overall to find malicious JPEG images. This work presents a malicious JPEG detector using machine learning. It extracts discriminative yet straightforward features from the JPEG file structure and leverages them with a machine learning classifier to discriminate between benign and malicious JPEG images. We evaluated malicious JPEGs extensively on a real-world representative collection of 156,818 images which contains 12,255 (91.44%) benign and 1805 (8.55%) malicious images. The results show that malicious JPEG, when used with the LightGBM classifier, demonstrates the highest detection capabilities, with an area under the receiver operating characteristic curve (AUC) of 95%, true positive rate (TPR) of 95%, and a very low false-positive rate (FPR) of 8%.

Keywords: JPEG, image, malware, detection, machine learning, features.

4.1 Introduction

Cyberattacks have increased a lot, targeting individuals, businesses, and organizations in the previous few years. Cyberattacks typically consist of unsafe actions such as stealing confidential information, spying or monitoring, and causing damage to the victim. Attackers may be aggravated by philosophy, unlawful intention, a wish for publicity, etc. Attackers continuously look for new and efficient methods to commence attacks and distribute a malicious payload. Documents or files shared by the Internet have frequently served to achieve this. As executable files (i.e., EXE) are identified as risky, attackers are progressively using non-executable files (e.g., PDF, Docx, etc.) that are wrongly measured to be harmless for many users. Some non-executable files permit an attacker to run arbitrary malicious code on the target machine when the file is opened. JPEG is the most well-liked image format. Almost all individuals, small and large organizations use JPEG on various platforms.

In November 2016, it was reported that attackers used Facebook Messenger to spread the infamous Locky ransomware via JPEG images. In August 2017, it was reported that SyncCrypt ransomware was spread using JPEG images. In December 2018, Trend Micro, an enterprise cybersecurity company, reported that cybercriminals used memes (JPEG images) on Twitter for conveying commands to malware. In December 2019, researchers from the Sophos security company published a comprehensive report on the MyKings cryptomining botnet that lurks behind a seemingly innocuous JPEG of Taylor Swift.

The ability to detect malicious JPEG images has great importance as individuals and businesses widely use JPEG images. Existing endpoint defense solutions which are based on signatures (e.g., antivirus) can only detect known malware. In contrast, in recent years, machine learning (ML) algorithms have demonstrated their ability to detect both known and unknown malware in various domains, particularly for detecting malware on multiple types of files. This work presents a machine learning-based solution for efficiently detecting unknown malicious JPEG images.

4.2 Background

In this section, we provide background material related to our research and technical information regarding the structure of a JPEG image. Since the JPEG file structure is complicated, we only present the basic information needed to enable the reader to comprehend and understand the proposed malicious JPEG solution offered in this research. The format of JPEG images is comprehensively described in the JPEG File Interchange Format (JFIF) specification.

4.2.1 JPEG file structure

JPEG stands for Joint Photographic Experts Group, which has become the most popular image format. In 1992, JPEG became an international standard for compressing digital still images. JPEG files usually have a filename extension of .jpg or .jpeg. A JPEG image file is a binary file which consists of a sequence of segments. Segments can contain other segments hierarchically. Each segment begins with a two-byte indicator called a "marker." The markers help divide the file into different segments. A marker's first byte is 0xFF (hexadecimal representation); the second byte may have any value except 0x00 and 0xFF. The marker indicates the type of data stored in the segment. Segment types are assigned names based on their definition or purpose; for example, 0xFFD9 is EOI, and 0xFFFE is COM. Segment types 0xFF01 and 0xFF0 -0xFF9 consist entirely of the two-byte marker; all other markers are followed by a two-byte integer indicating the size of the segment, followed by the payload data contained in the segment.

The very first marker we care about is FFD8. It tells us that this is the start of the image. If we do not see it, we can assume this is not a JPEG file. Another equally important marker is FFD9. It tells us that we have reached the end of an image file. Every marker, except FFD0 to FFD9 and FF01, is immediately followed by a length specifier, giving that marker segment's length. The image file start and end markers are always two bytes long each.

Figure 4.1 presents the possible markers, their hexadecimal code, and their definition/purpose. The work [38] provides more details about JPEG image, structure, each marker and its purpose.

4.3 Related Work

This research looks for related work that aims to detect malicious images and JPEG files using machine learning. However, we could not find any papers that apply machine learning to detect malicious JPEG images. It is important to emphasize that we only refer to images that contain malware or malicious code as malicious images. Therefore, the domain of adversarial image detection (e.g., [63], [30], [34], [45], and [23]) is different from this

```
        0    1    2    3    4    5    6    7    8    9    A    B    C    D    E    F
000:   FF   D8  FF   E0   00   10   .J   .F   .I   .F   00   01   01   01   00   48
010:   00   48   00   00  FF   DB   00   43   00   01   01   01   01   01   01   01
020:   01   01   01   01   01   01   01   01   01   01   01   01   01   01   01   01
030:   01   01   01   01   01   01   01   01   01   01   01   01   01   01   01   01
040:   01   01   01   01   01   01   01   01   01   01   01   01   01   01   01   01
050:   01   01   01   01   01   01   01   01  FF   DB   00   43   01   01   01   01
060:   01   01   01   01   01   01   01   01   01   01   01   01   01   01   01   01
070:   01   01   01   01   01   01   01   01   01   01   01   01   01   01   01   01
080:   01   01   01   01   01   01   01   01   01   01   01   01   01   01   01   01
090:   01   01   01   01   01   01   01   01   01   01   01   01  FF   C0
0A0:   00   11   08   00   02   00   06   03   01   22   00   02   11   01   03   11
0B0:   01  FF   C4   00   15   00   01   01   00   00   00   00   00   00   00   00
0C0:   00   00   00   00   00   00   00   09  FF   C4   00   19   10   01   00   02
0D0:   03   00   00   00   00   00   00   00   00   00   00   00   00   00   06   08
0E0:   38   88   B6  FF   C4   00   15   01   01   01   00   00   00   00   00   00
0F0:   00   00   00   00   00   00   00   00   07   0A  FF   C4   00   1C   11   00
100:   01   03   05   00   00   00   00   00   00   00   00   00   00   00   00   08
110:   00   07   B8   09   38   39   76   78  FF   DA   00   0C   03   01   00   02
120:   11   03   11   00  :86   F7   E7   1D   A9   16   CA   77   30   D0
130:   14   F7   41   DC   5A   8E   FB   31   19   26   5D   C4   2A   F4   5C   81
140:   7B   DB   06   84   A0   75   17  FF   D9
```

SEGMENTS — FIELDS — VALUES

- START OF IMAGE — marker — FFD8
- APPLICATION0 (DEFAULT HEADER) — marker/length FFE0/16; identifier JFIF\0; version 1.1; units 1 (dpi); density 72x72; thumbnail 0x0
- QUANTIZATION TABLE — marker/length FFDB/67; destination 0 (luminance); table (8x8) (100% quality)
- QUANTIZATION TABLE — marker/length FFDB/67; destination 1 (chrominance); table (8x8) (100% quality)
- START OF FRAME — marker/length FFC0/17; precision 8; line Nb; samples/line; components 3; 1d factor table 1 1x1 0 (LumY); 1d factor table 2 2x2 1 (ChromCb); 1d factor table 3 2x2 1 (ChromCr)
- HUFFMAN TABLE — marker/length FFC4/21; class 0 (DC); destination 0; 1 code of 1 bit 00; 1 code of 2 bits 09
- HUFFMAN TABLE — marker/length FFC4/25; class 0 (DC); destination 0; 1 code of 1 bit 00; 1 code of 3 bits 00 00; 1 code of 4 bits 38 88 B6
- HUFFMAN TABLE — marker/length FFC4/21; class 1 (AC); destination 0; 1 code of 1 bit 07; 1 code of 2 bits 0A
- HUFFMAN TABLE — marker/length FFC4/28; destination 0; 1 code of 3 bits 00 00 07 08; 1 code of 4 bits 00 38 39 76 78
- START OF SCAN — marker/length FFDA/12; components 3; selector / DC, AC table 1 0,0 / 1,1 / 1,1; spectral select 0..0); successive approx. 00
- IMAGE DATA (ENTROPY-CODED SEGMENT)
- END OF IMAGE — marker FFD9

Figure 4.1   JPEG image possible markers, their hexadecimal code, and their definition/purpose.

work's domain. The difference between adversarial images and malicious images lies in the part of the images that are altered to transform an original image into an adversarial or malicious image. Adversarial images are created by intentionally designing the pixels of the original image in such a way that a machine learning model will misclassify the image. In contrast, malicious JPEG images store the malicious mechanism in some metadata fields outside the pixels section; usually, the pixels are not changed to maintain the image's authenticity.

Kunwar et al. [44] proposed a theoretical framework that is aimed to detect the presence of data or code in JPEG images (without the use of machine learning). Their framework has three phases: steganography analysis, extraction of the embedded file, and uploading the extracted file to an online scanning tool such as VirusTotal or Metascan. However, the chapter does not describe any experiments performed to evaluate the framework or present any detection results on a real-world dataset.

In addition, we identified a study [47], which proposed an authentication method for JPEG images that can distinguish legitimate operations (e.g., compression) from malicious operations. However, "malicious" in the context of this work does not mean that the image carries malicious code as a payload.

Instead of that, the image is not authentic and has been manipulated. In addition, the work does not apply machine learning methods.

Most of the works on JPEG images focused on steganography methods, steganography analysis (steganalysis) methods, or adversarial images. However, this work focuses on the detection of malicious JPEG images. The solutions [54] and [17] provide approaches using machine learning methods for the detection of malicious JPEG images.

4.4 Methodology

In this section, we present the core working of our work, the machine learning-based solution for detecting malicious JPEG images. Malicious JPEG detector receives a JPEG image as input. The malicious JPEG detector pre-processes the image and extracts the planned features into a vector of features. The JPEG reader inspects the file statically and navigates through the JPEG image file structure to extract the features. The features are then transferred to a machine learning-based model, which outputs a classification (benign/malicious) for the input image. Figure 4.2 presents malicious JPEG architecture and solution building blocks. Malicious JPEG solution building blocks are as follows:

4.4.1 Input JPEG images

Our research uses around 12,255 JPEG images, including 11,207 benign files and 1048 JPEG files. This imbalance dataset has a 92:8 ratio for benign to

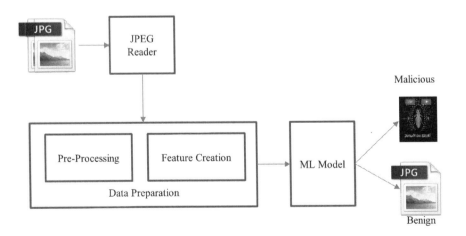

Figure 4.2 Malicious JPEG architecture.

malicious files, respectively, representing real-world scenarios. Images are collected from the IIT Kanpur corpus and can be extracted from VirusTotal. This module provides JPEG files to the next module JPEG reader.

4.4.2 JPEG reader

The JPEG reader is implemented in Python that inspects the JPEG files one by one statically, without viewing the image (which requires executing image viewer software that itself might have a vulnerability). The code traverses through the image file structure/markers from Start of Image (SoI) to End of Image (EoI) markers to extract information, i.e., Length of Marker and Image data size. Table 4.1 shows possible marker details and hexadecimal codes. This module generates around 33 features for each image. The author [38] provides a code snippet to extract markers information from JPEG images.

4.4.3 Data preparation

The data preparation phase performs the following actions:

- **Check outliers**–Checks distribution and make sure all numeric features are within limits.

Table 4.1   Possible marker/features from JPEG images.

Marker name	Hexadecimal code	Definition
APP_n	0xFFE0–0xFFEF	Reserved for application used
COM	0xFFFE	Comment
DAC	0xFFCC	Define arithmetic conditioning table(s)
DHP	0xFFDE	Define hierarchical progression
DHT	0xFFDC	Define Huffman table(s)
DNL	0xFFDC	Define number of lines
DQT	0xFFDB	Define quantization table(s)
DRI	0xFFDD	Define restart interval
EXP	0xFFDF	Expand reference image(s)
JPG	0xFFC8	Reserved for JPEG extensions
JPGn	0xFFF0–0xFFFD	Reserved for JPEG extensions
RES	0xFF02–0xFFBF	Reserved
RST_m	0xFFD0–0xFFD7	Restart with modulo 8 counter m
SOF_n	0xFFC0-3,5-7,9-B, D-F	Start of frame
SOS	0xFFDA	Start of scan
TEM	0xFF01	For temporary use in arithmetic coding
SOI	0xFFFD8	Start of Image
EOI	0xFFD9	End of Image

- **Exploratory analysis**–Univariate analysis and bivariate analysis to understand JPEG malware correlation with features.
- **Check multicollinearity and feature elimination**–Verify any same features using Pearson correlation and eliminate duplicate features.
- **One Hot Encoding**–Convert all categorical features to binary features using One Hot Encoding.
- **New features creation**–Apply different statistical aggregations on various features, such as the number of times each marker occurred, the minimum length of the marker, and the maximum length of the marker.

This module generates 60 features and target features that define the image as malware or not benign for all 12,255 images.

4.4.4 Machine learning model

This module split the image dataset into 70:20:10 ratios as train, test, and validation datasets. Train data is used to train, and test data is used to tune the training model during grid search. Validation data set is used to test fine-tuned model to check final performance.

The experiments utilize the following commonly used, high-performing classic and nonlinear machine learning classifiers: Random Forest, LightGBM, and Deep Learning MLP classifier. We select these classifiers as they perform well on highly imbalanced datasets. In our preliminary experiments, we examine classifiers from families other than the decision tree family, such as Logistic Regression and Naïve Bayes. However, they do not provide good results; therefore, we do not include them in our evaluation.

We applied the above-mentioned machine learning classifiers with Python using the following packages: **LightGBM, Random Forest, and MLP Classifier**. Grid search and cross-validation are used for hyperparameter tuning and validation of all classifiers. This module saves fine-tuned ML models for the prediction/detection of images.

4.5 Model Evaluation

4.5.1 Evaluation metrics

For evaluation purposes, we have calculated each classifier's true positive rate (TPR) and the false-positive rate (FPR). The TPR and FPR are the most important metrics in the domain of malware detection. A viable detection system must maintain a high TPR (representing the system's ability to detect

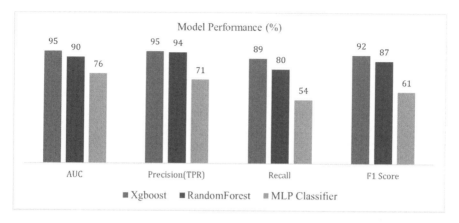

Figure 4.3   Model performance comparison.

positive samples (malicious) successfully) and a low FPR (the system's capability of avoiding false alarms for negative samples–benign). Note that since our dataset is highly imbalanced, it is essential to use the TPR and FPR metrics instead of the well-known accuracy metric. They represent the classifier's accuracy and false-positives for the minority class only, i.e., malicious. In addition, we measured the area under the receiver operating characteristic (ROC) curve, or the AUC, precision, recall, and F1-score of the machine learning classifiers trained and tested on different datasets. Figure 4.3 shows the model performance using various measures.

4.5.2 JPEG image detection

JPEG reader extracts features from JPEG images and pre-trained outputs the classification as benign or malicious. As shown in Figure 4.4 input image passes through the JPEG solution, which includes a JPEG reader that extracts features from JPEG images, and a saved pre-trained machine learning model

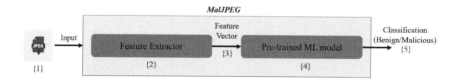

Figure 4.4   JPEG image detection.

is used to predict given input files as benign or malicious. Figure 4.4 shows the architecture of JPEG image detection.

4.6 Conclusion

Malicious JPEG, a machine learning-based solution, can detect unknown malicious JPEG images efficiently with 95% accuracy and only an 8% of false-positive rate. Best results are achieved using the LightGBM model. The AUC, TPR, TNR, and FPR are 95%, 93%, 92%, and 8%, respectively. This chapter presents JPEG, a machine learning-based solution for efficient detection of unknown malicious JPEG images. To the best of our knowledge, there is very little research to offer a machine learning-based solution explicitly tailored for detecting malicious JPEG images. Malicious JPEG extracts ten discriminative but straightforward features from the JPEG file structure. It leverages them with a machine learning classifier to discriminate between benign and malicious JPEG images.

4.7 Acknowledgments

We would like to thank IIT Kanpur for providing access to their malware collection for academic use and guidance for implementation.

5

Live Monitoring of Malware Attacks on Cloud using Windows Agent-based Solution

Sheetal A. Suvarna

E-mail: sheetalasuvarna@gmail.com

Abstract

Live monitoring malware for attacks in a highly regulated environment is imperative to prevent potential cybersecurity attacks. It requires many security levels to be deployed to analyze the behavior of more sophisticated malware. In modern days, there are hybrid methods for monitoring such malware by isolating virtual machines from the production environment or even deploying an agent-based solution. In many circumstances, the situation would be devastating to both the client and the organization; however, in many instances, the recovery seems minimal, and the loss of reputation is irrevocable. Furthermore, the frontline of defense needs to be built around a well-rounded system to circumvent the attacks. One can take arms with the sea of troubles stemming from malware attacks. In addition, robust defense-in-depth topology aids clear progression for malware analysis. In our work, we develop an agent for monitoring the malware attacks in near-real-time. This Windows agent-based malware monitoring on the cloud platform captures the snapshot. This analysis of signature-based anomalies provides security professionals insight into making informed decisions. The situation is far simpler to pick the open-source technologies and build the stack required to create the forerunner for malware classification and detection. The objective of the product is to build an agent-based open-stack product for malware classification and detection. Actionable research has shown Windows is one of the favourite places for malicious attackers, with over 83% of attacks last year. Another objective is to build on the cloud. The visual dashboard of the cynical process changes gets monitored on the production systems. It performs the following operations: capturing snapshots of all processes every 10 seconds, detecting and classifying potential malware based on its severity, and shipping the logs to the ELK stack for further analysis.

Keywords: Malware, malware analysis, anomalies, trojan, indicators of compromise, cuckoo, anti-evasion, sandbox, log analyzer, classification.

5.1 Introduction

Malware is a software that could be created by big business criminals purely to destroy files and make modest financial gains with an illegal intent. Parallelly, malware analysis is a study to understand the behavior, origin, and functionality while gathering the information without prior permission. All the while, malware development has become more sophisticated, with greater measures being taken by malware developers to check the authenticity of the operating environment before executing an attack, often described as context-aware malware or sandbox evasion in big, medium, and small businesses. We need to look at another set of averting, like IOC databases aid the malware analysts as they grapple with newer malware variants. In this work, we discuss the methodology and combination of open-source tools to detect and analyze potential malware along with visual representation on the cloud. It helps the SOC or security management team to interpret and take timely action to mitigate or avoid the risk depending on the severity of the damage a malware could cause to the environment.

5.1.1 About malware

There is a misguided assumption about malware that they are not a virus. It is a software regardless of how it works. Conversely, a virus is a specific type of malware that can self-replicate into different forms. Malware could be spyware, ransomware, etc., [81] that would damage the network or a targeted system. With the transformational epoch of digitalization, malware have been evolving since 1971. The maiden malware was "creeper," which was not created to damage the target, on the contrary, was used as a guessing game that moved from one system to another as an experiment, creation of ideated "reaper" to catch the creeper. There are some contentious discussions around which was the maiden malware and shall continue to argue its grounds.

5.1.2 Types of malware

- **Ransomware** – this disables the victim's access to the data until a ransom is paid.
- **Bots** – this could launch a large flood of attacks.
- **Worms** – this replicates itself and spreads to a particular network.
- **Adware** – this type flashes unwanted advertisements.
- **Spyware** – this collects the user behavior and the patterns without the knowledge of the user.
- **Trojans** – this is designed to damage, disrupt, steal, or in general, inflict some other harmful action on your data or network.
- **Mobile malware** – this one damages the mobile devices.
- **Keyloggers** – this monitors the keyloggers of the users without their knowledge.
- **Rootkits** – this gives attackers remote access to victim's system.
- **Fileless malware [43]** – this one alters the files which are native to the operating system.

5.1.3 Fileless malware

Persistent attacks on fileless malware are one of the latest versions of attacks. Its functionality attacks are based on a few parameters like the infection method, configuration data, injection data, and persistence method. It falls under the malware category of memory-resident malware such as SQL slammers, code red, poweliks, lurks, Windows registries malware like PowerWare and kovter, and rootkit, namely phase bot. The fileless malware can hide its location to make difficulties in the detection process by traditional AV solutions and the security analyst. The target could be both organizations or individuals. It is a rising trend to compromise a targeted system to avoid downloading malicious executable files, usually to disk; instead, it uses the capability of web exploits, macros, scripts, or trusted admin tools. For example, this could include an infection mechanism and legitimate system tools.

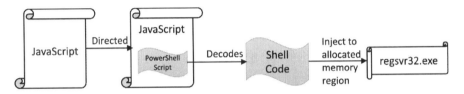

Figure 5.1   Fileless Javascript.

5.2 Background

Windows is the target of 83% of all malware attacks [69] because it is the most common operating system. Similarly, it has more than 50% of 660 security gaps in Windows 10. The extent of damage we have experienced in the few years is exponentially increasing with the pandemic situation since the online usage and the users have tremendously increased. The effort involved in the escape activities needs to be backed up both money- and skill-wise to grapple with the constant battles of red and blue hat professionals. At this rate, cybercriminals may create 160 million malware programs this year. Building a robust threat intelligence system is a kernel of truth. When this turmoil foments in the operating system, the damage shall be unfathomable to recover. There are 43 million new malware programs, which means 4.2 malware programs are developed every second. We must not subvert adversaries of Windows OS; instead, we take proactive measures to lessen the malware attacks and reduce the time to trigger the remediation strategies when it occurs. Windows operating system is vulnerable to the majority of the attacks and needs additional effort from the organization to keep the environment safe.

Therefore, there was a compelling reason to develop a Windows agent to extract information from the host which are as follows:

- Process and system calls
- Process tree
- File and registry
- Events and logs
- Process ID and name of the process
- System performance

Again a program should analyze the information captured by the agent, identify the suspicious activities based on the integrated IOC and classify its severity based on defined risk criteria. The summarized information could be presented to a security administrator in security operations center in a visual dashboard to further investigations.

Hindsight shows the considerable damages caused and soon can fan out based on various observations, recovery, and post facto programs arising from threat attacks. The Procmon utility helps to extract such information from the system to develop an agent for Windows.

5.2.1 Procmon

It comes with abundant advantages to monitoring the processes on the systems. It shows the real-time file system, and it is an advanced monitoring tool indispensably used by millions, if not zillions. Procmon stems from Windows Sysinternals utilities and legacy Filemon and Regmon, and it is part of the Microsoft TechNet website. Broadly, it is used to track the system and any software activity, precisely to track the process or application that accesses a registry key or a file.

- **Procmon operations:** It brings a lot of individuality to perform operations which are as follows:
  - Advanced logging architecture scales to a prodigious extent,
  - setting non-disclosure filters helps in avoiding data loss;
  - Image path, session ID, command lines can be identified effortlessly,
  - The root cause of the operation can be easily identified with the thread stacks,
  - Process tree displays all the inter-connected processes to quick traceability.

  Installation of this agent would bring unstoppable awesomeness to a highly regulated environment or systems while being ready for malware attacks.

- **Features of Procmon:** It has two legacy Sysinternals utilities, Filemon and Regmon. It also contains extensive enhancements, including rich and non-destructive filtering, comprehensive event properties such as session IDs and user names, reliable process information, logging to a file, and full thread stacks with integrated symbol support for each operation. This tool is used as a malware hunting toolkit. This tool is fundamentally open-source, and its Windows Sysinternals utility is both powerful and inexpensive. Integration with IOC databases and VirusTotal will create a strong influential network for an in-depth analysis of large data processing servers.

5.3 Project Approach

The flow of our work is shown in Figure 5.3. This project is dissected into four parts, the first being the detection and then analysis, followed by classification and reporting. Figure 5.2 shows the milestones of our work. The four phases are as follows:

- Detection engine – Agent development
- Analysis engine
- Classification engine
- Reporting with ELK

5.3.1 Detection engine – agent development

The flow chart in Figure 5.2 depicts the entire process flow to gain the following objectives of the product. The infinite loop ensures real-time monitoring, and the timer of 10 seconds keeps comparing with the previous snapshot against the signature-based anomalies. The log event captures any process changes, Regkey, and File systems like file modification or directory modification. This log file is then mapped with all the system events to output a final log for the analysis phase. In the analysis phase, potential malware and its severity are provided for actionable decision-making.

Figure 5.4 shows the working of the agent developed. Figure 5.5 shows the overview of the detection engine. Windows agent installed on the host machine captures real-time events at the kernel level using process monitor and detects the possible anomalies in the client machine by capturing the processes snapshot. It tracks the incorrect permissions on a file registry key, registry keys or values that are missing or even misnamed, and the required application if missing. We can view registry, file system, process and network activity in the Process Monitor tool. The agent uses Procmon and other native utilities to take snapshots every 10 seconds, capturing all mercurial anomalies and eventually shipping them to the cloud platform to trigger the analysis engine.

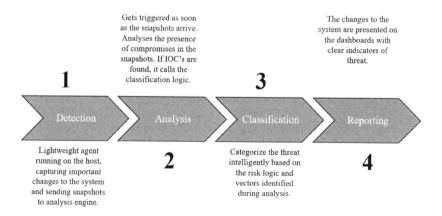

Figure 5.2   Project Milestone.

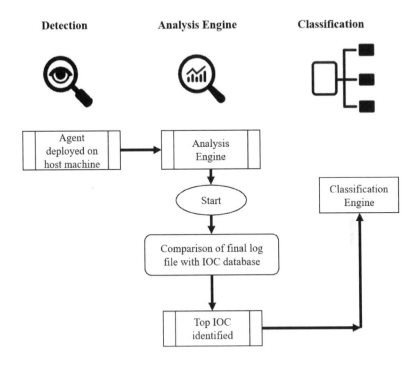

Figure 5.3 Product flow chart.

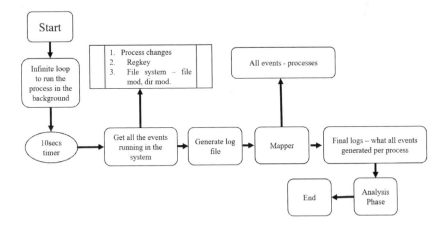

Figure 5.4 Windows agent using Procmon analyzing the events.

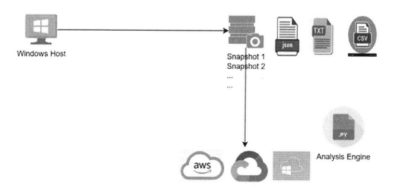

Figure 5.5   Detection phase overview.

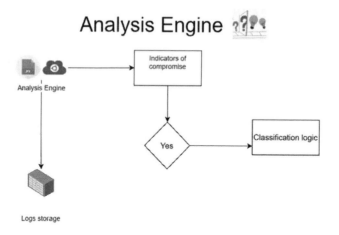

Figure 5.6   Analysis overview.

5.3.2 Analysis engine

Figure 5.6 depicts the overall structure of the analysis engine working. The analysis process is initiated based on the time limit that is set on the agent, and there are two methods for this approach which are as follow:

- Replacing the snapshots when a new snapshot is taken, by doing this, not only the space required would be lesser but also the prediction of the process to zero down the actual process that caused the attack, especially for the last process that was executed.
- Creating multiple snapshots with a criterion whether last 10 or 20 snapshots, these snapshots are available in the form of the CSV files. This solution provides elaborate information to identify the source or even the time the malware was introduced. The analysis engine analyzes data and stores data against IOCs defined upon identifying in moves to the next phase for classification.

IOC analysis engine determines whether the data is malicious or not. After submitting the data, the threat level is assessed with the processes running on the windows machine. Comparatively, the time is taken to identify the threat, and the response may be drastically reduced with the IOC analysis. This stellar contribution to the community and the IOC database would be beyond reproach.

```
1  from io import StringIO
2  import pandas as pd
3  import sys
4
5  def read_iocfile(path, delimiter = None):
6      with open(path) as readfiles:
7          if delimiter:
8              return readfiles.read().split(delimiter)
9          return readfiles.read().splitlines()
10     # search_string = []
11     # if delimiter == 'nl':
12     #     with open(path, 'r') as readlines:
13     #         for i in readlines:
14     #             if i != '\n':
15     #                 search_string.append(i)
16     # if delimiter == ',':
17     #     with open(path,'r') as readlines:
18     #         for i in readlines:
19     #             for item in i.split(','):
20     #                 if item != '\n':
21     #                     search_string.append(item)
22     # return search_string
23
24 def csv_analysis(input):
25
```

```
26  #C:\\Python_project\\csv_compare\\drive-download-20210730
       T053152Z-001\\IOC_rootmalware.txt

27

28      backdoorIOClist = read_iocfile('C:\\Python_project\\
       csv_compare\\drive-download-20210730T053152Z-001\\
       bckdor_fulacs_procIOC.txt',',')

29      adwareBorwserIOClist = read_iocfile('C:\\Python_project
       \\csv_compare\\drive-download-20210730T053152Z-001\\
       adware_browser.txt',',')

30      dilers_procIOClist= read_iocfile(C:\\Python_project\\
       csv_compare\\drive-download-20210730T053152Z-001\\
       dilers_procIOC.txt',',')

31      dwlodrs_spwr_ProcIOC = read_iocfile('C:\\Python_project
       \\csv_compare\\drive-download-20210730T053152Z-001\\
       dwlodrs_spwr_ProcIOC.txt',',')

32      trojanIOC = read_iocfile('C:\\Python_project\\
       csv_compare\\drive-download-20210730T053152Z-001\\
       trojanIOC.txt',',')

33      IOC_rootmalware= read_iocfile('C:\\Python_project\\
       csv_compare\\drive-download-20210730T053152Z-001\\
       IOC_rootmalware.txt')

34      IOC_userlevel = read_iocfile('C:\\Python_project\\
       csv_compare\\drive-download-20210730T053152Z-001\\
       IOC_userlevel.txt')

35

36      # print(backdoorIOClist)
37      # print(adwareBorwserIOClist)
38      # print(IOC_userlevel)
39      # print(IOC_rootmalware)
40      # print(trojanIOC)
```

```
1   #/home/sheetal/Desktop/Logs/
2       path = sys.argv[1]

3

4       #file_path = [os.path.abspath(p1) for p1 in os.listdir(
       path)]
5       #csvfiles = glob.glob('C:\\Python_project\\csv_compare
       \\*.csv')
6       # input = requests.get('http://localhost:8000/snapshot.
       csv')
7       print(type(input.content))

8

9       df = pd.read_csv(StringIO(input.content.decode()))

10

11      df['adwareBorwserIOClist']=df['Process Name'].isin(
       adwareBorwserIOClist)
```

```
12    df['dilers_procIOClist']=df['Process Name'].isin(
      dilers_procIOClist)
13    df['dwlodrs_spwr_ProcIOC']=df['Process Name'].isin(
      dwlodrs_spwr_ProcIOC)
14    df['trojanIOC']=df['Process Name'].isin(trojanIOC)
15    df['root_priv']=df['Path'].isin(IOC_rootmalware)
16    df['IOC_userlevel']=df['Path'].isin(IOC_userlevel)
17
18    # for files in csvfiles:
19    #     df = pd.read_csv(files)
20    #     df['adwareBorwserIOClist']=df['Process Name'].isin
      (adwareBorwserIOClist)
21    #     df['dilers_procIOClist']=df['Process Name'].isin(
      dilers_procIOClist)
22    #     df['dwlodrs_spwr_ProcIOC']=df['Process Name'].isin
      (dwlodrs_spwr_ProcIOC)
23    #     df['trojanIOC']=df['Process Name'].isin(trojanIOC)
24    #     df['root_priv']=df['Path'].isin(IOC_rootmalware)
25    #     df['IOC_userlevel']=df['Path'].isin(IOC_userlevel)
26
27
28    df.to_csv('final.csv')
```

```
1  import csv
2  import glob
3  import os
4  import pandas as pd
5
6  def write_to_csv(row_to_write):
7      with open('final.csv', 'a') as output_file:
8          csv.writer(output_file).writerow(row_to_write)
9
10
11
12 def check_regkey(check_csvstring, ioc_list):
13     for search_string in ioc_list:
14         if search_string in check_csvstring:
15             return True
16     return False
17
18
19 def check_IOCs(csv_matching_col, search_string_list):
20     # with open(csv_matching_col, 'r') as readline:
21     for line in csv_matching_col:
22         for search_string in search_string_list:
23             if search_string in line:
```

```
24                    return True
25       return False
26
27 def read_iocfile(path,delimiter):
28     search_string = []
29     if delimiter == 'nl':
30         with open(path,'r') as readlines:
31             for i in readlines:
32                 if i != '\n':
33                     search_string.append(i)
34     if delimiter == ',':
35         with open(path,'r') as readlines:
36             for i in readlines:
37                 for item in i.split(','):
38                     if item != '\n':
39                         search_string.append(item)
40     return search_string
41
42 #//home//sheetal//regKey_IOCs//IOC_rootmalware.txt
43
44 backdoorIOClist = read_iocfile('//home//sheetal//regKey_IOCs
       //bckdor_fulacs_procIOC.txt',',')
45 adwareBorwserIOClist = read_iocfile('//home//sheetal//
       regKey_IOCs//adware_browser.txt',',')
46 dilers_procIOClist= read_iocfile('//home//sheetal//
       regKey_IOCs//dilers_procIOC.txt',',')
47 dwlodrs_spwr_ProcIOC = read_iocfile('//home//sheetal//
       regKey_IOCs//dwlodrs_spwr_ProcIOC.txt',',')
48 IOC_rootmalware= read_iocfile('//home//sheetal//regKey_IOCs
       //IOC_rootmalware.txt','nl')
49 IOC_userlevel = read_iocfile('//home//sheetal//regKey_IOCs//
       IOC_userlevel.txt',',')
50
51 print(IOC_rootmalware)
52
53
54 #/home/sheetal/Desktop/Logs/
55 path = '/home/sheetal/Desktop/Logs/snapshot4.csv'
56
57 #file_path = [os.path.abspath(p1) for p1 in os.listdir(path)
       ]
58 csvfiles = glob.glob('/home/sheetal/Desktop/Logs/*.csv')
59 #csvfiles =
60 #print(file_path)
61 for files in csvfiles:
62     df = pd.read_csv(files)
```

```
63    # print(df)
```

```python
1  import os
2  from flask import Flask, request, redirect, url_for,
       send_from_directory
3  from werkzeug.utils import secure_filename
4  import Compare_IOC_analysis
5
6  UPLOAD_FOLDER ='upload'
7  ALLOWED_EXT = 'csv'
8
9  app = Flask(__name__)
10
11 def allowed_fileext(filename):
12     return filename[-3:].lower() in ALLOWED_EXT
13
14 @app.route('/', methods=['GET', 'POST'])
15 def upload_file():
16     if request.method =='POST':
17         file = request.files['files']
18         if file and allowed_fileext(file.filename):
19             print('**found file', file.filename)
20             filename = secure_filename(file.filename)
21             file.save(os.path.join(app.config['UPLOAD_FOLDER
   '], filename))
22             return url_for('uploaded_file', filename=
   filename)
23         return
24     '''
25     <!doctype html>
26     <title>Upload new FIle</title>
27     <h1>Upload new File</h1>
28     <form action="" method=POST enctype=multipart/form-data>
29         <p><input type=file name=file>
30             <input type=submit value=Upload>
31     </form>
32     '''
33
34 @app.route('/uploads/<filename>')
35 def uploaded_file(filename):
36     return send_from_directory(app.config['UPLOAD_FOLDER'],
37                                 filename)
38 if __name__ == '__main__':
39     app.run(host = '0.0.0.0', port=5001,debug=True)
```

5.3.3 Classification engine

The theme is the built-in logic and vectors identified in the analysis phase in this phase. They result from the IOC sample database used for the project. In the absence of classification, the question becomes curly to resolve the attack. Many problems stem from the way classification logic is defined. If there are many false positives, security analysts are busy fixing the false-positives, which means the actual vectors lose their importance. If managed well, classification could reduce redundant efforts for security analysts. We define various labels for the attacks based on different scenarios; these are as follows:

1. **High** – an indicator of critical action the malware is executing, like connecting to the command center.
2. **Medium** – an action that can adversely affect the host or the network.
3. **Low** – an indicator of less intrusive but malicious activity by the malware.
4. **Suspicious** – defined for generic criteria for suspicious files.
5. **Malicious** – file or code delivered over the network to infect the host.
6. **Dangerous** – program or file that may cause severe damage to the computer intentionally.

Classification acts as an ability to recognize the earlier detected similar features between the processes. The abstract group would have to be segregated according to their behavior, signature, and traits. Perhaps, this is the most less altruistic advantage to focus the available time for the security focal to remediate on recovery strategies or the extent of the research required to reduce the impact of such malicious activities. A few approaches to reduce the malicious activities are defined as isolation, confinement, jail, disconnecting from the production environment or redirecting to a different server, especially the vaccine approach to simulate the upheaval substance in production servers. With varying levels of success, classification profoundly impacts the post facto strategies for the organization.

5.3.4 Reporting with ELK

Figure 5.7 shows the overview of the reporting phase. Reporting helped in providing insights into the right priorities at the right time. At the onset, it offers essential details that could be used to develop future forecasts and improve decision-making. Besides, regular reporting enables businesses to measure and compare potential malware and increase their own customized IOC list that could be added to the existing resources. To clarify, periodical

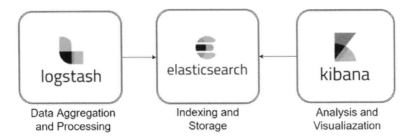

Figure 5.7 Reporting engine overview.

reporting produces massive hindsight for future enhancements, and mainly they help channel the investments on the business-critical mission.

Elasticsearch [48] is an analytical engine based on Apache Lucene. It is an open-source search engine. It is used as a tool for log management and analytics for indexing and storing data. It is armed with HTTP RESTful API, enabling accelerated search in almost real-time. It is developed on Java and supports other languages like Python, C#, Ruby, and PHP.

Logstash is an open-source log aggregator and processor that works by reading data from many sources and sending it to one or more destinations for storage or stashing – in this case, when using ELK for data analytics, to Elasticsearch. This data is massaged and processed, and well-shaped for a structural view. It comes with ready-to-use filters, codecs, and can process extensive data.

While Kibana is also an open-source analysis and visualization engine that works on top of Elasticsearch and Logstash, this is the most preferred tool for visualization. It helps ease the complexity of data into meaningful patterns and trend analysis. Although there may be many other tools, ELK is not just a buzzword, it is a well-known and preferred tool with numerous plugins too!

5.4 Deployment Architecture

The proposed approach assures continued motivation, securing and achieving the goal, and meeting the deadline without compromising the quality of work. There is no statistical illusion but statistical absoluteness. It has been put through the test, and the economic values this could bring to the organization to improve security standards are enormous.

5.4.1 Product tool architecture (benefits of the agent)

Product tool architecture denotes the product features and the relationship between features and functionalities. The schematic representation of the Windows agent monitors local services. It then reports any issues, and this can be installed on multiple machines as it displays real-time file system and process/thread activity registry files. Several benefits of this agent are as follows:

- High-quality software
- Light-weight agent
- No lock-in with vendors
- Offline payload analysis
- Scaling and abundance support
- Regulatory and compliance

5.5 Product Future Enhancements

Cloud deployment for such a complex environment will provide additional insights for SOC team analysis and response based on the server deployment size and geographical spread. Docker image of the agent, detection, analysis, classification, and respective reporting engines are the first set; also, applying machine learning models will provide higher accuracy to refine the greater prediction precision. UI (user interface)based parameters allow customization, depending on the risk appetite. In Parallel, Lambda functions and events queues in the cloud-native architecture reduce cost and gives a reliable system for ease of management. Figure 5.8 shows the idea about the future enhancement plan.

For example, Azure is the cloud-native architecture. The Windows agent can be installed in cloud virtual machines to monitor the host Windows servers or Windows endpoints. Azure file storage can be used for the event logs storage, with the Azure machine learning engine accuracy can be improvised. Multiple clusters can be deployed in the Azure function like analysis, classification, and

Figure 5.8   Future enhancements.

reporting clusters as shown in Figure 5.9. Cloud deployment provides greater customisation and faster response to nascent but scaling threats.

5.6 Conclusion and Future Directions

In conclusion, predicting and preparing for malware attacks is wiser to combat the production system carnage that comes along with malware attacks. As the digital realm becomes ever more entwined with the physical, there has been a growing trend for a military-style lexicon concerning cyberattack. It has a technological step-change that can significantly influence and change the world of work as they knew before our proposed work. In our work, Windows agent is a lightweight open-source software and could be deployed on both Windows server and Windows endpoint. Since this is highly flexible, it can be integrated with many IOC databases for identifying anomalies. It would be best suited for the highly regulated environment that processes sensitive personal information (SPI). The insights this agent could provide cannot be trivialized. For many attackers, the easiest path forward is running out of the runway, and as forerunners, we need to analyze the trends and patterns constantly.

Experts consider a large-scale retrofit of malware IOC's signature database to provide a forerunner for further improvements for the ever-evolving malware attacks. Understanding the markets for future malware trends depicts the most vulnerable health care, financial sectors, and supply chain attacks, where threat actors will continue to leverage vendors and sub-contractors. Assuming the presumes of historical data, deploying robust malware prevention tools avoids

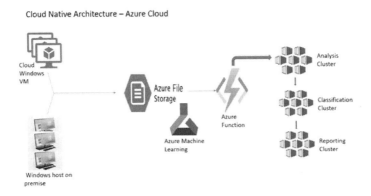

Figure 5.9 Future enhancements.

large financial loss and goodwill for any-sized organization. It is found that malevolent activities could initially be developing, but scaling, the newer the variants, the remediation could be just a band-aid solution.

5.7 Acknowledgments

I would like to thank Professor Sandeep Shukla, Professor Rohit Negi, Professor Anand Handa, Professor Venkatesan, in collaboration with Indian Institute of Technology, Kanpur, C3i Center and learning partner TalentSprint for approving this project and encouraging me to study this subject.

6

Malware also Needs "Attention"

**Atharv Singh Patlan, Som Vishwas Tambe, Yathartha Goswami,
Nitesh Kumar, Anand Handa, and Sandeep K. Shukla**

C3i Center, Indian Institute of Technology, Kanpur, India
E-mail: atharvsp@iitk.ac.in; somvt@iitk.ac.in; niteshkr@cse.iitk.ac.in;
anand@c3ihub.iitk.ac.in; sandeeps@cse.iitk.ac.in

Abstract

Malware continues to pose a deeply evolving challenge to the world of security. There is an ongoing fight between attackers and malware analysts. Traditional malware detection methods require a lot of time and human resources. We turn to machine learning-based solutions for the problem in the hand of identifying malicious programs. In this chapter, we analyze the malicious properties present within the API call sequence patterns of the programs. We use API fragments and LSTM-based model with attention layers for classification. We present our experimental results on two publically available datasets. Our method based on API fragments and techniques like attention gives better performance than other works that adopt similar techniques after comparing the experiments.

Keywords: Malware, machine learning, natural language processing.

6.1 Introduction

There has been a lot of research on malware analysis and classification of malware. This topic is of interest among many researchers, and various tools and techniques are developed for it. There already exists a lot of literature on tackling the problem of malware detection and malware classification. However, the malware remains a severe problem to individuals, companies and organisations as attackers continuously use it as a tool to get confidential information or perform attackers on the other machine. Malware analysis can be performed using static or dynamic analysis techniques. Though static analysis techniques are powerful and accurate but the attackers hide the program's main intent through techniques like obfuscation, which leads to failing most of the static analysis techniques. The attackers are becoming clever daily and have even deployed techniques like polymorphism to reorder the codes and create multiple virus variants. This demands for developments of techniques that are less cumbersome and more adaptable to changes in the programs.

Deep learning has seen a significant rise in almost every field like image processing, audio recognition, language translation and whatnot. Even seemingly unrelated fields like software engineering have started deploying these techniques nowadays. Since machine learning has the remarkable ability to facilitate the task of feature extraction from low-level data, many scholars have naturally resorted to machine learning techniques for detecting malware. In studies like [41], people have used image processing techniques to classify malware. The authors in [76] used API call sequences to detect malware. In [55] uses signal processing techniques and NLP methods to handle the assembly code and building a model using LSTMs. These methods have definitely shown the effectiveness of deep learning in this field, but they still suffered from being unstable and getting easily disturbed.

We chose to study the programs dynamically and hence extract out the behavior of the program during execution. This chapter explored this problem using a machine learning perspective and tried different techniques for the same. Here, we have explored the idea to combine techniques of NLP with malware analysis.

Our method is used on programs made for the Windows platform by extracting the API execution sequences. To exploit the local malicious properties present in a program, we cut the API sequence into smaller API fragments and worked on them. We generate word embeddings for many API calls in our approach and then use these embeddings to generate sequences (sentences) of these calls. We use the analogy from English vocabulary and practice frameworks like Word2Vec to prepare the required embeddings. Using such a technique allows us to generate semantically valid embeddings for each API call, which naturally helped us in the process employed ahead. We use two layers of LSTMs and two layers of attention in our model. Since an API call can be highly correlated to a previous API call in a different fragment, so we use the attention layers to help us model the relation between API calls present in different fragments. This embedding represents the program and operates as an input for the machine learning classifiers. The chapter also discusses the further classification of the program into the type of malware class using binary classification.

We are experimenting on multiple datasets it indicates that our technique is stable and produces good results even with only a few thousand samples in hand. The approach used in this chapter of combining LSTMs with attention network outperforms the other works that use similar strategies. Our approach is stable toward techniques like obfuscation since we are using API calls at the very heart of our proposed approach. Furthermore, using the model on a binary for prediction is also relatively easy since it just involves extraction of API sequences which can be automated using analysis systems like Cuckoo Sandbox [31].

Our work contributes as follows:

- Use the analogy of language vocabularies and using Word2Vec like models to generate embeddings that made sense semantically and naturally helped achieve good results.
- Analyze the local malicious properties by converting them into fixed-length API fragments.
- Continuing with the analogy with language, we used NLP models like LSTMs to locally utilize the features/knowledge present in the API sequences.
- Combining the normal LSTMs with attention layers to get the correlations present between calls globally.
- Our work demonstrates the effectiveness of LSTM models and techniques like attention in malware analysis which can be taken up and explored further for research.

The paper is structured into the following sections further. Section 6.3 discusses the current literature present for this field, Section 6.4 explains the proposed model in detail and provides theoretical significance of the method, Section 6.5 provides experimental evidence to validate our idea, and finally, Section 6.6 concludes the chapter.

6.2 Related Work

It has been quite a few ages since the start of the era of the development of malware. Nowadays, attackers have become more clever in preparing these malicious programs, and there is endless competition between malware developers and malware analysts. The speed of malware development is relatively high, and everyday malware developers come up with new and more sophisticated ways. The malware being developed in recent times is highly complex and uses obfuscation techniques, which makes it extremely difficult to analyze such programs. Malware can be analyzed using either behavior-based or signature-based techniques. The signature-based techniques though being fast, lose their effectiveness against obfuscation techniques and the behavior-based techniques since they require observing the behavior take a lot of time and make the task much more cumbersome for the analyst. Thus, the detection of malware using traditional approaches like heuristic-based, graph-based, entropy-based, etc., is not possible. To tackle this ever-growing field of malware development, analysts need to learn from the malware program's behavior. Thus, machine learning provides a solution to develop classification models to get ahead of the new variants of malware.

Malware analysis techniques fall mainly under these two broad categories.

- *Static malware analysis*
  In this analysis, various static features of a program like hash values, opcodes, strings, and PE header information are extracted without executing the program. The malicious programs are disassembled using tools like IDA Pro, Capstone, Radare, etc. and then assembly code is examined to get the execution flow of the file and patterns present in the file for detecting some signs of malicious activity. This type of analysis suffers from the drawback of being highly time-consuming and much more complicated. Techniques deployed by developers like obfuscation such as code encryption, reordering instructions, dead code insertion further make the analyst's task much harder.

- *Dynamic malware analysis*
  In this technique, the malware is executed on the host system by making virtual environments and then logging the program's activities. Various activities of the program like file system operation, process generation and execution, API calls, and network activities are observed. Based on them, the file is classified as benign or malicious. The files which can not be analyzed using static analysis techniques can be analyzed using this technique.

Since the traditional methods suffer from the requirement of a significant workforce and time, we turn to machine learning-based approaches. Machine learning-based methods are highly generalizable and do not require much manual work. Machine learning can even learn some features that are too difficult or can not be extracted manually because of its ability to learn. In [76], the authors use the assembly file of the program (produced by the disassembler) and convert the assembly bytecodes into pixel features and then use CNNs to learn. Although this deploys the program information, an attacker can still confuse the classifier by inserting external assembly functions. The authors in [84] use SVM to build a malware detection framework based on the concept of supervised learning and achieve good results. In [7], the authors use the API calls appearing made by the program. Their method relies on a single malicious API that could appear on a series of call sequences, and only the exact API sequence is harmful.

Many researchers use the graph-based analysis techniques [18][37][61]. The authors in [37] use graph-based techniques as well as deep learning techniques for malware analysis. The authors in [15] use the concept of clustering based on a given binary's family dependency graph. The authors in [65] use deep learning to create embeddings for malware based on their API call graphs. In [82], the authors use high-level features of extracted behavior graph using stacked autoencoders. This method is precise and has the disadvantage of working on the whole sample while malicious fragments are only partial, which affects the prediction of malicious behavior.

The authors in [74] analyze the API attributes and propose a map color method based on categories and occurrence time for a unit time the API executed and then use CNNs for classification. In [83], the authors proposed a new statistical method that is based on extra information addition and removal and hence led to reduction of the length of API call sequences. These sequences are then fed to LSTMs for training. Researchers also explore ways to extract features using machine learning techniques. In [6], the authors explore ways of extracting features from the frequency of API and compare them with other neural networks. The methods based on API call sequences are accurate. Still, they suffer from long execution sequences appearing in the program while the actual malicious part is a tiny portion of the total code. The method of extracting efficient sequences is explored in [46], but they only retain the sequential nature of the code execution. In [50], authors tackle the problem of finding similarity between two binary functions by producing the function embeddings through a self-attentive neural network. They also provided ways to detect malware by comparing the given program with a program known to be malicious.

6.3 Proposed Methodology

6.3.1 Datasets

We use namely two datasets in our experiments. The first dataset is collected from Oliveira *et al.* [13]. It consists of 42,797 malware API call sequences and 1079 benign API call sequences. We use this to train our network for binary classification by splitting the dataset into 1079 examples of malware and benign classes, which we split in a 7:3 train to test ratio. The dataset consists of the first 100 API call sequences of various program files created for running in the Windows Operating System.

The second dataset we use is collected from Catak *et al.* [60] which features 7107 malware API call sequences across eight categories, namely *Trojan, Backdoor, Downloader, Worms, Spyware Adware, Dropper, Virus*. We use this dataset for our eight-way classification experiments. The dataset consists of unfiltered API call sequences of various malicious programs created for attacking the Windows Operating System. The API call sequences are of diverse lengths, varying from a length of just 10 to a maximum of 400K API calls. The sequences also consist of repeating calls.

Due to the almost similar natures of all the categories from the execution standpoint and the wide variety in API calls, we consider Dataset 2 a more challenging dataset than Dataset 1.

6.3.2 Methodology

The methodology pursued by us aims to employ the ideas of natural language processing to the problem of malware detection, as highlighted by us earlier, and we make use of API calls used by malware to perform this task.

Our main motivation for using this approach is the similarities we find in how the API calls are arranged in any binary file and how the words are arranged in any document in a natural language.

For example, consider any sentence from a document in the English language and any function from a Windows executable. Just like the sentence is composed of words, the function is composed of API calls. Furthermore, consider the natural language dependency graph [14] of a sentence in the English language such as *"I saw the boy who lives here"* displayed in the Figure 6.1. The structure and connections of the generated graph are very similar to the structure of the API call graph generated from API call sequences, which is shown in Figure 6.2. It is intuitive as different API calls in the call graph should be related in purpose and operate upon the results of the previous API call, similarly as words in a sentence are related and operate upon the context till the previous word.

Thus we decided to capture this intuition of modeling API call sequences as a human language with the help of natural language processing techniques. Our approach first tries to judge the functionality of different API calls by

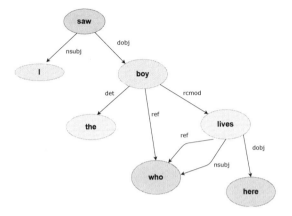

Figure 6.1 Sentence dependency graph in english.

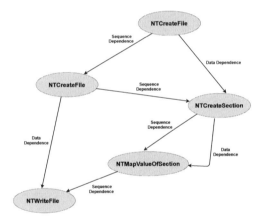

Figure 6.2 An API call graph. As can be seen, its structure is very similar to that in Figure 6.1.

finding related API calls in the entire corpus, like building up a vocabulary of words.

Subsequently, we establish relations at a function and binary file-level (sentence and document level for a language). That is, we use methods to train our classifier such that it understands what different API calls (words) represent when presented in a specific order to form a function (sentence). Further, it needs to understand how the various functions (sentences) combine to form the binary file (document). We employ Word2Vec and LSTMs with an attention mechanism.

6.3.2.1 API call level understanding – Word2Vec

To get a semantic understanding of the API call sequences, we use Word2Vec [52]. Word2Vec is a popular method to generate meaningful representations and understand the semantics of words and sentences and use it mainly for natural languages by working on their inherent properties and structures. It figures out the relation between different words and observing the similarities in the structure of natural language sentences and API call sequences. We apply the same to generate embeddings for various API calls and thus use it to perform malware analysis.

The usage of Word2Vec allows us to measure the relationship between different API calls at a word level in terms of a natural language.

6.3.2.2 Function and binary file level understanding – attention

To understand what the API calls and their corresponding functions represent when they occur in a particular order. It is necessary to understand the context in which they occur because it is possible that the same set of instructions when arranged in one order, lead to a harmless file (which might not work correctly) and dangerous malware. It is very similar to human languages, where words and sentences, when arranged in different orders, lead to different meanings.

We make use of attention [75] for this purpose of taking into consideration the context of the different API calls and functions, along with the LSTM layers. It offers a performance gain and helps in highlighting parts of sentences that are of higher relevance and learn the semantics irrespective of the order in which words, or phrases, occur.

We showcase our intention through Table 6.1. The first column shows a sentence in the English language, and the words highlighted in [29] according to their attention score related to the task given. We give three more examples from our dataset [13] and the tasks being the functionality of the API calls. We take three tasks – *Cryptography*, *System Metrics*, and *Resource Handling*. The sentences are the 10-length sequences picked up directly from the dataset. For *Cryptography*, *CryptAcquireContextW*, *CryptCreateHash*, and *CryptHashData* are all examples of functions which are related to a cryptograhic module which performs operations for authentication, encoding, or encryption. In the *System Metrics* task, *GetSystemMetrics* is the API function from Windows User Controls header file *Winuser.h* which provides system metrics, for example, the width of a cursor in pixels, or the number of display monitors on a desktop. The *Resource Handling* task also accurately

Table 6.1 Attention visualization for API call sequence. The first column gives an example of a sentence in English.

Task: Cleanliness	Task: Cryptography	Task: System metrics	Task: Resource handling
Not	CryptAcquireContextW	GetSystemMetrics	LoadResource
the	NtOpenKey	NtClose	DrawTextExW
cleanest	NtQueryValueKey	GetSystemMetrics	GetSystemMetrics
rooms	NtClose	NtAllocateVirtualMemory	FindResourceExW
but	NtOpenKey	LdrLoadDll	LoadResource
bed	NtClose	LdrGetProcedureAddress	GetSystemMetrics
and	LdrGetProcedureAddress	LdrGetDllHandle	DrawTextExW
bathroom	CryptCreateHash	FindResourceExW	LdrGetDllHandle
was	LdrGetProcedureAddress	LoadResource	FindResourceExW
clean	CryptHashData	FindResourceExW	LoadResource

highlights the *LoadResource* and *FindResourceExW* functions which are used to handle resources in the memory.

Firstly, to get a representation of the relationship between the API calls at a function level, we model the functions as separate sentences and feed them into an LSTM with an attention layer on top. The output is a vector representation of a function which is formed by applying self-attention on the constituent API calls of the function.

It is to be noted that the API call sequences obtained are usually wildly varying in length and have a lot of repetitions, which makes it difficult for them to be modeled as ordinary length sentences as found in human language. Thus we preprocess the API calls by allowing only a set number of consecutive repetitions of the same API call.

Further, we used the *N*-gram model and considered fixed-length sequences of these API calls equivalent to one sentence of a human language. *N*-grams constitute words, which makes *N*-grams sentences.

Subsequently, to establish a relationship between the different functions and thus in the entire Binary file, we use a second LSTM layer with attention, which works similarly to the previous LSTM layer, however, working with vectors representing functions and not individual API calls. The output of these two layers is a vector representation of the entire binary file, which considers the order and contexts of its constituent API calls and functions. Thus, it is an accurate representation of the properties of the file, much like the embeddings produced by NLP models when a human language document is passed through them.

6.3.3 Network architecture

We train a custom Word2Vec model, as mentioned earlier. Word2Vec helps us in getting embeddings for a completely new vocabulary of API call sequences. We train the Word2Vec model on Dataset 1[13]. Figure 6.3 shows a small snippet of an API call sequence. We present two API calls, namely *SetFilePointerEx* and *StartServiceV*, and the top 10 closest API calls calculated by measuring the cosine similarity between the embeddings generated by our trained Word2Vec model. As observed in both the examples indicated in Figure 6.3, the functions found closest are quite related to the ones being compared. For instance, *StartServiceW* is an API function to start a service, and the two most similar API functions found are *OpenServiceW* and *OpenSCManagerW*. These functions open an existing service, establish a connection to the service control manager, and open the specified service control database. Similarly, *SetFilePointerEx* also gives us closely related functions such as *NTWriteFile* which writes data to an open file, and *SetEndOfFile* which sets the physical file size for the specified file to the current position of the file pointer. It verifies the fact that Word2Vec successfully train a model which identifies the semantics between API calls.

In total, we experiment with two varieties of models. One of them is the vanilla flavor, which contains two stacked LSTM layers with two linear layers. In the second one, we attach an attention layer after each LSTM layer, which we hypothesize will enable the representations to incorporate more relevant and dominant API calls/sequences in the vector-space representations. This will be passed to further layers to give us more meaningful and accurate representations, hence delivering better results in identifying newer and fresh malware.

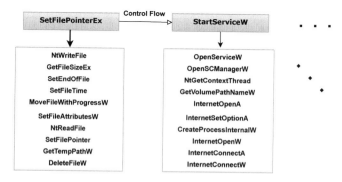

Figure 6.3   API call sequences, and their respective 10 closest API functions.

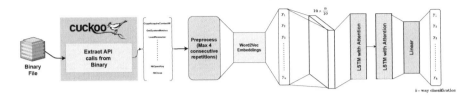

Figure 6.4 Pipeline for our model.

With the above setup, we conduct experiments for both binary and eight-way classification.

Furthermore, our model consists of an embedding layer that shares the weights of the custom-trained Word2Vec model, from which we use the embeddings $v \in \mathbb{R}^k$. As mentioned above, we experiment with models consisting of two stacked LSTM layers. The model with attention has single attention layers succeeding each of the LSTM layers, after which follows a batch normalization layer and two dense layers, which are common in both of the models above.

Figure 6.4 shows the pipeline of our approach in which one may automate the process using Cuckoo Sandbox before preprocess (Max 4 consecutive repetitions) block. After the preprocess (Max 4 consecutive repetitions) block, one may automate all the steps using our approach.

6.4 Experiments and Results

6.4.1 Experimental setup

For the conduction of experiments, we use two variants of our model architecture as described in Network Architecture subsection of Proposed Methodology section. We use a dropout rate of 0.2 in both our LSTM layers and an N-gram size of 10 for our experiments. The length for N-gram was chosen for reasons as referred to in [40]. We use the Adam optimizer with default configuration for training. The word embedding size is set to 20, as this value is giving us the maximum cosine similarity between related API calls.

We conduct experiments by first freezing the Word2Vec embeddings, pre-training the deep layers, and then unfreezing the Word2Vec layers for fine-tuning to learn more robust features. In case any unknown API calls are encountered in the input, they are given a $< UNK >$ tag and are assigned embeddings equal to the average of all embeddings corresponding to the known API calls.

Table 6.2 Results of experiments on Dataset 2 [60] on various models. The best results on each malware type are in **bold**. The precision of the results is upto two decimal places for consistency with the reported results in the literature.

Model	Adware	Backdoor	Downloader	Dropper	Spyware	Trojan	Virus	Worm	Average
Adaboost [28]	0.76	0.52	0.69	0.57	0.41	0.51	0.74	0.57	0.60
Decision Tree [11]	0.45	0.40	0.51	0.37	0.11	0.16	0.41	0.78	0.40
kNN [77]	0.70	0.57	0.67	0.45	0.32	0.32	0.62	0.49	0.52
RF [12]	0.48	0.62	0.52	0.35	0.17	0.16	0.80	0.58	0.42
2-layer LSTM [35]	0.77	0.56	0.59	0.44	0.42	0.28	0.68	0.45	0.52
Ours (no Attention)	0.84	0.77	0.83	0.84	0.80	0.71	0.92	0.82	0.82
Ours (Attention)	**0.94**	**0.85**	**0.96**	**0.87**	**0.84**	**0.77**	**0.96**	**0.88**	**0.88**

In order to conduct ablation studies and showcase the advantages of using attention for malware classification, we also present results without using the attention mechanism after the LSTM layers. We perform our experiments, modeling them as classification problems. For Dataset 1 [13], we classify the samples as either malware or benign.

Similarly, we report the results using Dataset 2 [60]. Even this dataset is a multi-class dataset, and we report the results as a binary classification problem – performing classification as class 1 to be Trojan and 0 for other class once, then in the next iteration, classifying the samples as class 1 to be Backdoor and 0 for rest, and so on. We report the results in this fashion to be consistent with the existing literature using this dataset and the relatively small quantity of samples of each class available.

Evaluation metrics: The datasets we use are mostly unbalanced and generally lean more toward one category than the other. For example, for Dataset 2, as we are modeling it as a binary classification problem, the number of samples with label 0 is approximately seven times that of those with label 0. Thus along with accuracy, we need to report class-wise metrics such as Recall to display our model's robustness and classification abilities.

We also report the same metrics for Dataset 1 along with the Precision and F1-score. However, we use the same number of test samples for both classes in this dataset to maintain consistency with results in the literature.

6.4.2 Results

6.4.2.1 Dataset 1

It is necessary to make the number of training and testing samples equal for Dataset 1, which is highly imbalanced, as described in Datasets subsection to perform experiments. Thus, we randomly sample 1079 data points from

Table 6.3 Results of experiments on Dataset 1 [13] on various models. The best results for each metric are in **bold**.

Method	F1-score	Precision	Recall	Accuracy
1-Layer DGCNN [13]	0.9076	0.8879	0.9283	0.9105
2-Layer DGCNN [13]	0.9201	0.9216	0.9186	0.9244
LSTM [35]	0.8738	0.8542	0.8932	0.8727
Our approach (no Attention)	0.9586	0.9508	0.9667	0.9583
Our approach (Attention)	**0.9697**	**0.9586**	**0.9810**	**0.9693**

malware, take all the 1079 data points from the benign category, and then randomly divide these in the ratio of 7:3 in the train to test data. The remaining samples from malware class are discarded as they result in an imbalance in training. This exact procedure is followed by the original paper presenting the dataset, [13], which ensures experimental consistency.

We compare the results of both methods with and without attention with the results presented by [13], that is, using 1 and 2 layer deep graph convolutional neural networks and a two-layer LSTM. The results using various evaluation metrics are shown in Table 6.3.

As see in the results present in Table 6.3, our methods far outperform the results of the methods reported in [13]. We also report the accuracy of a two-layer LSTM method. We feed-forward the outputs from the first LSTM to the second without using any N-grams and concatenating them before being passed as input in the second LSTM layer. The results on a single-layer LSTM come out to be even lower than the DGCNN results. It shows that using just the recurrent properties of LSTM is not enough to ensure enough attention paid to the inputs and the context in the API calls is taken care of by the network.

Our results are also visibly better when using attention with our method, as compared to without it. It shows that using just Word2Vec embeddings is not enough to ensure that the network utilizes the context-dependence of the API calls. We need a dedicated attention mechanism to ensure that the context is utilized.

6.4.2.2 Dataset 2

Dataset 2 is relatively complicated, as mentioned in Datasets subsection. The API calls in this dataset are repeating in nature and also are varying significantly in length. Thus, it is important to preprocess this dataset.

We observe that the significant variation in lengths from 10 to 400K is mostly due to repetitions. If we removed any two consecutive calls to the same API function, it resulted in a maximum length of 345 API calls. Thus, to test the effectiveness of our formulation, we allowed a small amount of repetition in which we let a maximum of four consecutive calls to the same API function to get our final data points. In contrast, excessive repetitions are removed, and the first distinct function call took its place after all the repeats.

Subsequently, we obtain a dataset where the maximum length of the input is 485, while the shortest length stays at 10. In order to ensure good results and make sure that this high variation in lengths does not cause a problem in classification, we trim the API call sequences to a maximum length of 200.

Following these preprocessing methods allow us to get our final dataset to conduct experiments. We again use this dataset in a binary classification setting for comparing with other results available in the literature. Thus to perform experiments on one class, we labeled all the samples belonging to that class as 1 and the remaining samples as 0. We use an 8:2 ratio train-test split, ensuring that the number of samples with label 1 and 0 are equal in the training and testing data to demonstrate our performance better.

We compare the results of both of our methods with the results present in [60], which uses a simple two-layer feedforward LSTM, learning the embeddings while training. We also compare our results with popular machine learning algorithms which do not utilize deep feature extraction layers, using TF-IDF vectors as the embeddings. The results are presented in Table 6.2. We report the class-wise accuracy as it is a multi-class dataset.

As we see in Table 6.2, both our methods perform much better than the other methods in all the tasks and have a significant boost in the mean performance per class. Our methods also perform very well on the harder classification tasks in this dataset: Spyware and Trojan (which have significantly less recall values when the other methods are used).

Similar to Table 6.3, our attention-based method performs better than the method without attention, even in Table 6.2.

6.5 Conclusion

In this work, we explore a way for analyzing the maliciousness in the program using the API call sequences present in that. We utilize the inherent structure of these API call graphs by looking at their similarity with the dependency graphs of the English language. It hints us to go for the domain of natural language processing naturally.

Hence, we design an NLP-based detection framework for the detection of malicious programs. Modeling the API segments in the English language sentences and then learning the features using techniques like "Attention" and NLP models like LSTMs help us produce better results than previous work adopting similar methods. The experimental results also show that our approach is stable and efficient across different datasets. Our work effectively explores the application of NLP-based techniques in the malware analysis field, which can have important significance on future researches in this field. In future work, one may explore more complex NLP-based primitives like transformers instead of LSTMs and deploying the method as a practical application is also a potential future work after optimizing the technique and pipeline further.

Part III

IDS

7

Implementation of an Intrusion Detection System and Deception Technologies using Open Source Tools for Small Businesses

Purushartha Srivastava and Kalpesh Seludkar

E-mail: purusharthasrivastava1993@gmail.com; kseludkar13@gmail.com

Abstract

Small businesses are considered the backbone of any country's economy. Therefore, keeping functioning the business is very crucial to ensure economic growth. We all know how today's business giants, Google, Amazon, and Apple, started as small startups and where they are today. Today, almost all businesses are being run and operated through the Internet. It exposes a business to various vulnerabilities and threats. A study says that small businesses fall victim to ransomware attacks every 14 seconds, and 60% of small businesses go out of business within six months of a cyber Attack. So, what if a business goes out of business at the startup phase? We might lose the next business giant. Our proposed model is to implement the network and host-based intrusion detection system and deception technologies, i.e., Honeypots. All implementations are done using open-source tools to provide cost-efficient but effective solutions to small businesses who either do not have a cybersecurity budget or lack cybersecurity knowledge and expertise. We deployed Cowrie Honeypot, WordPress Honeypot, Mailoney Honeypot, and Dionaea Honeypot to publish the common services on the Internet. We deployed Suricata network intrusion detection system (NIDS) for network traffic analysis and network-based intrusion detection. Wazuh is deployed on each honeypot to detect host-specific threats and misconfigurations. A centralized management server (CMS) is built on Elasticsearch Cluster that indexes and stores all the logs from all honeypots and forwards them to Wazuh Manager for event correlation and trigger alerts in case of any cyberattack or anomaly detected.

7.1 Introduction

The Internet is growing fast in terms of users, data, and connected devices, and this rapid growth has increased people's dependency on the Internet. There are around 200 billion devices connected to the Internet in the present cyber world, and the number is expected to increase as remote work has become more commonplace after the COVID-19 pandemic. It increases the attack surface and opens opportunities for hackers and ransomware, which means that cybersecurity has become the most important than ever in today's Internet world. The number of cyberattacks varies from day to day and is increasing consistently.

It is hard to build proper controls and monitoring mechanisms to prevent all cyberattacks. All large-and medium-scale companies usually have dedicated budgets and resources for cybersecurity, but small businesses consider this as an unnecessary cost until a cyberattack impacts them.

Technologies evolve day by day as similar attackers apply attacks with new strategies. We always try to find the intention of the attacker and what strategies/tools they employ to attack systems. Knowing the attacker's intentions and strategies can mitigate the threat at the earliest. All the gathered information always helps to prevent attacks and protect the business.

This work describes how such small-scale businesses can deploy different honeypots resembling their technology stacks for gathering and analyzing threat intelligence data. This study provides insight into the honeypot deployment on the cloud with different geolocation, building real-time analytics capabilities and using HIDS and NIDS. It will help to analyze the real-time attacks on our honeypots and enhance the cybersecurity of the production system.

7.2 Tool Setup and Architecture

This section describes the high-level architecture of our work. It includes an introduction to the tools and technology we have used in this work. We have used draw.io [3] application to draw our project architecture diagram. Draw.io is a free diagramming application to draw diagrams and create flowcharts within the browser. It has a rich set of shapes, including software, servers, networking shapes, and Icons.

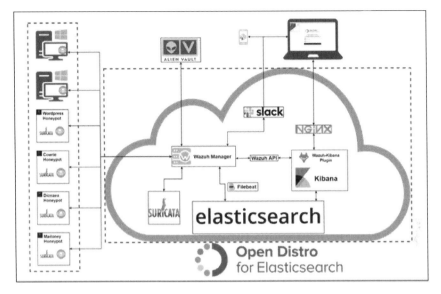

Figure 7.1 Architecture of proposed work.

We used Digital Ocean [36] cloud platform to host central management server (CMS) and honeypot servers. Digital Ocean is a cloud infrastructure provider that allows to create virtual private servers (also known as droplets). Each droplet is an individual virtual machine (VM) which is easy to deploy, manage, and scalable. We utilized the droplets to host different honeypots at different data centers located in different locations globally, though easy to manage from a single web-based UI. Digital Ocean platform has SSD-based virtual machines to optimize the performance at the most affordable cost. It is possible because of the hourly and monthly payment option, which helps manage the expense based on the requirement.

Our work architecture is divided into two parts – central management server (CMS) and endpoints or honeypots, as highlighted in Figure 7.1.

7.2.1 Central management server (CMS)

The central management server (CMS) is the main component of this project. It has several sub-components, such as the Elasticsearch, the centralized data storage for all logs from all endpoints and honeypots forwarded by the Wazuh agent. Wazuh manager collects the logs and performs normalization and enrichment on logs. The filebeat agent then transports the enriched logs to Elasticsearch for indexing. Kibana is the web interface to view the logs and create visualization and dashboard for better understanding. The wazuh manager performs event correlation and threat intelligence and triggers an alert when it detects an anomaly or cyberattack, which is then forwarded to the Slack notification. Various CMS components are as follows:

7.2.1.1 OpenDistro for Elasticsearch

Open Distro for Elasticsearch [8] is an open-source tool which combines the OSS distributions of Elasticsearch and Kibana with a large number of open-source plugins. These plugins play an important role in the OSS distributions.

7.2.1.2 Wazuh manager

The wazuh manager analyzes the data received from honeypots and endpoints, processes events through decoders and rules, and uses threat intelligence to look for well-known IOCs (indicators of compromise). A single wazuh manager analyzes data from hundreds or thousands of endpoints and can be scaled when set up in cluster mode. The manager is also used to manage the agents, configuring and upgrading them remotely when necessary. Additionally, the server can send commands or instructions to the agents to trigger a response when a threat is detected. There are two different options for deploying wazuh which are as follows:

- **All-in-one deployment:** Wazuh and Open Distro for Elasticsearch are installed on the same host.
- **Distributed deployment:** Each component is installed on a separate host as a single-node or multi-node cluster. This type of deployment provides high availability and scalability of the product, and it is convenient for large working environments.

7.2.1.3 Suricata

We know that most security issues are successfully detected by inspecting a server's network traffic. It is where a NIDS (Network intrusion detection system) can provide additional insight into the network [72]. Suricata is one such NIDS solution, which is open-source and can be quickly deployed. Because Suricata can generate JSON logs of NIDS events, it integrates perfectly with Wazuh. We installed Suricata on CMS and all honeypot servers to capture live network traffic and analyze the traffic for cyberattack.

7.2.1.4 Nginx reverse proxy

A reverse proxy works on behalf of a server, intercepting traffic and routing it to a separate server. There are several reasons you might want to install a reverse proxy. One of the main reasons is privacy. A reverse proxy can help balance loads between servers and improve performance if you have multiple servers. As a reverse proxy provides a single point of contact for clients, it can centralize logging and report across various servers. Nginx can improve performance by serving static content quickly and passing dynamic content requests to Apache servers [66].

7.2.2 Endpoints or honeypots

A honeypot is a server configured to detect an intruder by mirroring a real production system. It is configured as an ordinary server doing work, but this is a trap for an attacker. Honeypots are the best way to learn more about how the attacker exploits different services. Capture and monitor attacker activity in an isolated environment and gather information about attacker tactics and techniques. There are two types of honeypots, high-interaction and low-interaction. These types are defined based on the honeypot system's protocol services or interaction level when the attacker interacts with the system as they would any regular server operating system, to capture much information about the attacker's techniques, known as high-interaction honeypots.

In low-interaction honeypots, we emulated services with a limited subset of the functionality attacker would expect from a server, with the intent of detecting sources of a hacker's activity. We capture a minimal amount of data in low-interaction honeypots. We have deployed various honeypots in this work which are as follows:

7.2.2.1 Cowrie honeypot

Cowrie honeypot is implemented to emulate a vulnerable SSH and Telnet server. This honeypot is implemented to capture the attacker's technique which they applied to access the SSH server.

7.2.2.2 WordPress honeypot

WordPress honeypot invites attackers to perform an attack on the WordPress website configured on LAMP. It detects probes for plugins, themes, and other common files used to fingerprint a WordPress installation. We include the following component in WordPress honeypot:

- **Web Server Apache:** The Apache HTTP server is an open-source web server that is cross-platform software and freely available on the Internet, released under Apache License 2.0. It comes under Apache Software Foundation, and it is developed and maintained by an open community of developers. The Apache HTTP server is the most popular HTTP client on the web. Apache is a component of the most common web application stacks, LAMP or Linux, Apache, MySQL, and PHP, which are used as a web server, and this is needed for a web application stack to deliver web content.
- **Database MySQL:** MySQL is also the most popular component of LAMP, and it is an open-source relational database management system (RDBMS). A database stores data into one or more data tables and set-up the relationship between tables, such as one-to-one, one-to-many relations. We use relational databases to maintain a large amount of data and access the expected records. Structure query language is used to create, update, and fetch the records from relation databases and control users to access relational databases.
- **PHP:** PHP stands for Hypertext Preprocessor. PHP is one of the most widely used server-side scripting languages for web application development. Popular websites like Facebook, Wikipedia, Yahoo, etc. It is a scripting language used to develop static and dynamic web pages and applications.

7.2.2.3 Honeypot dionea

Dionaea honeypot is implemented to emulate vulnerable services like FTP, HTTP, SMB, etc., capture the attacker's behavior and collect binaries transferred to the server by intruders.

7.2.2.4 Honeypot mailoney

Mailoney honeypot is used for the SMTP service, written in Python. Some modules or types provide custom modes to fit your needs, like open_relay, postfix_creds, and schizo_open_relay.

- **open_relay:** We will attempt to log full-text emails attempted to be sent, it is just an open relay.
- **Postfix_creds:** This module logs credentials from login creds attempts.
- **schizo_open_relay:** This module logs everything and supports the functionality of open relay and postcreds.

We have configured two Windows 10 machines to capture and analyze the logs. We have installed Wazuh agent on both systems to forward all logs to Wazuh manager. It will include Windows event logs, service logs, network logs, scan installed applications, and Windows update logs.

7.2.2.5 Wazuh agent

The Wazuh lightweight agent is designed to perform several tasks with the objective of detecting threats and, when necessary, trigger automatic responses. The agent's core capabilities are as follows:

- Log and events data collection
- File and registry keys integrity monitoring
- Inventory of running processes and installed applications
- Monitoring of open ports and network configuration
- Detection of rootkits or malware artifacts
- Configuration assessment and policy monitoring
- Execution of active responses

The Wazuh agents run on many different platforms, including Windows, Linux, Mac OS X, AIX, Solaris, and HP-UX. They can be configured and managed from the Wazuh server.

7.3 Implementation of Tools

In this section, we see the implementation of individual components we discussed in the tool architecture Section 7.2 and how we integrated the components to detect real-world intrusions.

7.3.1 Create droplet on digital ocean

We created droplets on a DigitalOcean cloud platform for every honeypot server and CMS. By following the steps, one can create DigitalOcean droplet.

- **Logging in to DigitalOcean:** Sign up or login to DigitalOcean platform using the link https://cloud.digitalocean.com/login. Once logged in, go to DigitalOcean dashboard, then click on Create button to initiate droplet creation. The dashboard is shown in Figure 7.2.
- **Choosing an image:** Next, we choose an image of the "64-bit Ubuntu 16.04" OS for CMS and honeypots servers both.

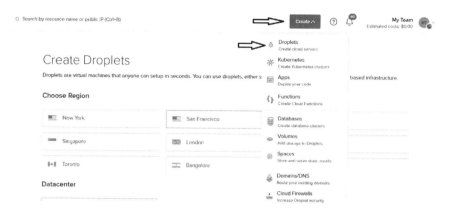

Figure 7.2   DigitalOcean dashboard.

- **Choosing a size:** Then, we choose the amount of RAM and storage space according to our need. We used 16 GB RAM and 320 GB storage images for CMS.
- **Adding backups:** Then, we can automatically enable backups and a snapshot of the live system, and this image can restore the existing droplet or create a new one according to need.
- **Choosing a data center region:** We choose the data center reason.
- **Select SSH keys:** We use SSH key-based authentication, which provides more security and convenience.
- **Finalizing and creating:** We provide any hostname to your newly created droplet [20].

7.3.2 Deploying Wazuh with open Distro for Elasticsearch

This section shows how to install Elasticsearch, Wazuh manager, Filebeat, and Kibana on the newly created droplet.

Installing Wazuh manager: The Wazuh installation on CMS starts with adding the Wazuh repository and installing the necessary packages for the installation [21].

```
1 # apt install curl apt-transport-https unzip wget libcap2-
      bin software-properties-common lsb-release gnupg
2
3 Install the GPG key:
4 # curl -s https://packages.wazuh.com/key/GPG-KEY-WAZUH | apt
      -key add -
```

```
 5
 6 Add the repository:
 7 # echo "deb https://packages.wazuh.com/4.x/apt/ stable main"
   | tee -a /etc/apt/sources.list.d/wazuh.list
 8
 9 Update the package information:
10 # apt-get update
11
12 Install the Wazuh manager package:
13 # apt-get install wazuh-manager
14
15 Enable and start the Wazuh manager service:
16 # systemctl daemon-reload
17 # systemctl enable wazuh-manager
18 # systemctl start wazuh-manager
```

Installing Elasticsearch:

```
1 Install Elasticsearch OSS and Open Distro for Elasticsearch:
2 # apt install elasticsearch-oss opendistroforelasticsearch
```

Configuring Elastic search:

```
1 Run the following command to download the configuration file
  /etc/elasticsearch/elasticsearch.yml:
2 # curl -so /etc/elasticsearch/elasticsearch.yml https://
  packages.wazuh.com/resources/4.2/open-distro/
  elasticsearch/7.x/elasticsearch_all_in_one.yml
```

Elasticsearch users and roles: running the commands will add users in Wazuh Kibana plugin.

```
1 Run the following commands to add the Wazuh users and roles
  in Kibana:
2 # curl -so /usr/share/elasticsearch/plugins/
  opendistro_security/securityconfig/roles.yml https://
  packages.wazuh.com/resources/4.2/open-distro/
  elasticsearch/roles/roles.yml
3
4 # curl -so /usr/share/elasticsearch/plugins/
  opendistro_security/securityconfig/roles_mapping.yml
  https://packages.wazuh.com/resources/4.2/open-distro/
  elasticsearch/roles/roles_mapping.yml
5
6 # curl -so /usr/share/elasticsearch/plugins/
  opendistro_security/securityconfig/internal_users.yml
  https://packages.wazuh.com/resources/4.2/open-distro/
  elasticsearch/roles/internal_users.yml
```

- **wazuh_user**: It is created for users who need read-only access to the Wazuh Kibana plugin.
- **wazuh_admin**: It is recommended for users who need administrative privileges.

These users created by running the above commands can be used to log in to Kibana console, but they are protected, and modification cannot be done from the Kibana console. One must run a security admin script to modify or add new users or roles.

Certificates creation: Remove the demo certificates.

```
1  # rm /etc/elasticsearch/esnode-key.pem /etc/
        elasticsearch/esnode.pem /etc/elasticsearch/kirk-key.
        pem /etc/elasticsearch/kirk.pem /etc/elasticsearch/
        root-ca.pem -f
2
3  # curl -so ~/wazuh-cert-tool.sh https://packages.wazuh.
        com/resources/4.2/open-distro/tools/certificate-
        utility/wazuh-cert-tool.sh
4
5  # curl -so ~/instances.yml https://packages.wazuh.com/
        resources/4.2/open-distro/tools/certificate-utility/
        instances_aio.yml
6
7  Run the wazuh-cert-tool.sh to create the certificates:
8  #   bash ~/wazuh-cert-tool.sh
9
10 Move the Elasticsearch certificates to their
        corresponding location:
11 # mkdir /etc/elasticsearch/certs/
12 # mv ~/certs/elasticsearch* /etc/elasticsearch/certs/
13 # mv ~/certs/admin* /etc/elasticsearch/certs/
14 # cp ~/certs/root-ca* /etc/elasticsearch/certs/
15
16 Enable and start the Elasticsearch service:
17 # systemctl daemon-reload
18 # systemctl enable elasticsearch
19 # systemctl start elasticsearch
20
21 Run the Elasticsearch securityadmin script to load the
        new certificates information and start the cluster:
```

```
22 # export JAVA_HOME=/usr/share/elasticsearch/jdk/ && /usr
     /share/elasticsearch/plugins/opendistro_security/
     tools/securityadmin.sh -cd /usr/share/elasticsearch/
     plugins/opendistro_security/securityconfig/ -nhnv -
     cacert /etc/elasticsearch/certs/root-ca.pem -cert /
     etc/elasticsearch/certs/admin.pem -key /etc/
     elasticsearch/certs/admin-key.pem
23
```

24 Run the following command to ensure that the installation is successful:
```
25 # curl -XGET https://localhost:9200 -u admin:admin -k
26
```
27 An example response should look as follows:
28 Output
```
29 {
30 "name" : "node-1",
31 "cluster_name" : "elasticsearch",
32 "cluster_uuid" : "tWYgqpgdRz6fGN8gH11flw",
33 "version" : {
34 "number" : "7.10.2",
35 "build_flavor" : "oss",
36 "build_type" : "rpm",
37 "build_hash" : "747e1cc71def077253878a59143c1f785afa92b9
      ",
38 "build_date" : "2021-01-13T00:42:12.435326Z",
39 "build_snapshot" : false,
40 "lucene_version" : "8.7.0",
41 "minimum_wire_compatibility_version" : "6.8.0",
42 "minimum_index_compatibility_version" : "6.0.0-beta1"
43 },
44 "tagline" : "You Know, for Search"
45 }
```

7.3.3 Installing Filebeat

Filebeat is the central management server tool that forwards alerts and archived events to Elasticsearch Cluster.

1 Install the Filebeat package:
```
2 # apt-get install filebeat
3
```
4 Download the preconfigured Filebeat configuration file:
```
5 # curl -so /etc/filebeat/filebeat.yml https://packages.wazuh
      .com/resources/4.2/open-distro/filebeat/7.x/
      filebeat_all_in_one.yml
```

```
 6
 7 Download the alerts template for Elasticsearch:
 8 # curl -so /etc/filebeat/wazuh-template.json https://raw.
     githubusercontent.com/wazuh/wazuh/4.2/extensions/
     elasticsearch/7.x/wazuh-template.json
 9 # chmod go+r /etc/filebeat/wazuh-template.json
10
11 Download the Wazuh module for Filebeat:
12 # curl -s https://packages.wazuh.com/4.x/filebeat/wazuh-
     filebeat-0.1.tar.gz | tar -xvz -C /usr/share/filebeat/
     module
13
14 Copy the Elasticsearch certificates into /etc/filebeat/certs
     :
15 # mkdir /etc/filebeat/certs
16 # cp ~/certs/root-ca.pem /etc/filebeat/certs/
17 # mv ~/certs/filebeat* /etc/filebeat/certs/
18
19 Enable and start the Filebeat service:
20 # systemctl daemon-reload
21 # systemctl enable filebeat
22 # systemctl start filebeat
23
24 To ensure that Filebeat is successfully installed, run the
     following command:
25 # filebeat test output
26
27 An example response should look as follows:
28 Output
29 elasticsearch: https://127.0.0.1:9200...
30 parse url... OK
31 connection...
32 parse host... OK
33 dns lookup... OK
34 addresses: 127.0.0.1
35 dial up... OK
36 TLS...
37 security: server's certificate chain verification is enabled
38 handshake... OK
39 TLS version: TLSv1.3
40 dial up... OK
41 talk to server... OK
42 version: 7.10.2
```

7.3.4 Installing Kibana

Kibana is a web interface for visualizing the events, alerts, and archives events.

```
1  To install the Kibana package:
2  # apt-get install opendistroforelasticsearch-kibana
3
4  Download the Kibana configuration file:
5  # curl -so /etc/kibana/kibana.yml https://packages.wazuh.com
       /resources/4.2/open-distro/kibana/7.x/kibana_all_in_one.
       yml
```

In the /etc/kibana/kibana.yml file, the setting server.host has a default value of 0.0.0.0, which means the Kibana interface can be accessed from anywhere outside. This value can be changed for a specific IP. In our case, we have changed it to 69.55.54.251.

```
1  Create the /usr/share/kibana/data directory:
2  # mkdir /usr/share/kibana/data
3  # chown -R kibana:kibana /usr/share/kibana/data
```

Now, install the Wazuh Kibana plugin. The installation of the plugin must be done from the Kibana home directory as follows:

```
1  # cd /usr/share/kibana
2  # sudo -u kibana /usr/share/kibana/bin/kibana-plugin install
       https://packages.wazuh.com/4.x/ui/kibana/wazuh_kibana
       -4.2.3_7.10.2-1.zip
3
4  Copy the Elasticsearch certificates into /etc/kibana/certs:
5  # mkdir /etc/kibana/certs
6  # cp ~/certs/root-ca.pem /etc/kibana/certs/
7  # mv ~/certs/kibana* /etc/kibana/certs/
8  # chown kibana:kibana /etc/kibana/certs/*
9
10 Enable and start the Kibana service:
11 # systemctl daemon-reload
12 # systemctl enable kibana
13 # systemctl start kibana
```

7.3.5 Installing Nginx as a reverse proxy

In this section, we will see how to install and configure the Nginx server as a reverse proxy [57].

```
1  To install Nginx from Default Repositories:
2  # sudo apt-get update
```

```
 3 # sudo apt-get install nginx
 4
 5 Once the installation completed, to start Nginx:
 6 # sudo systemctl start nginx
 7
 8 And to configure it to start at the time system boot, enable
      Nginx service using:
 9 # sudo systemctl enable nginx
10
11 To check Nginx is running:
12 # sudo systemctl status nginx
13
14 Now we will configure it as a Reverse Proxy.
15
16 First, we need to unlink the Default Configuration File. To
      do that:
17 # sudo unlink /etc/nginx/sites-enabled/default
18
19 create a new configuration file:
20 # sudo vi /etc/nginx/sites-available/reverse_proxy.conf
21
22 Edit the configuration file with any editor and add below
      configuration block. We will ensure to redirect HTTP
      traffic to HTTPS
23
24 server {
25 listen 80 default_server;
26 server_name _;
27 return 301 https://$host$request_uri;
28 }
29 server {
30 listen 443;
31 ssl on;
32 ssl_certificate          /etc/nginx/cert.crt;
33 ssl_certificate_key      /etc/nginx/cert.key;
34 access_log      /var/log/nginx/reverse-access.log;
35 error_log       /var/log/nginx/reverse-error.log;
36 location / {
37 proxy_pass https://69.55.54.251:5601;
38 }
39 }
40
41 Now, we will link and activate the new Configuration File.
      To do that, run:
42 ln -s /etc/nginx/sites-available/reverse_proxy.conf /etc/
      nginx/sites-enabled/reverse_proxy.conf
```

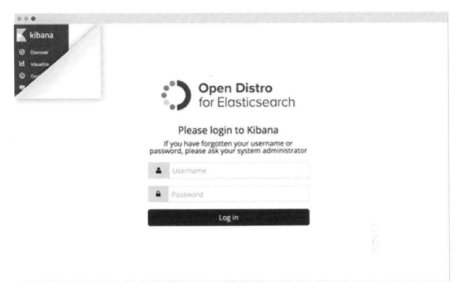

Figure 7.3 Kibana login console.

```
43
44 Test the new configuration:
45 # sudo service nginx configtest
46
47 Restart the NGINX service:
48 # sudo service nginx restart
```

Once the Nginx service is started, now you can access the Kibana console over https from your browser at https://69.55.54.251 as shown in Figure 7.3.

One can now log in to Kibana console with the wazuh_user or wazuh_admin user we have created during the Installing Elasticsearch section.

7.3.6 Installing Suricata on CMS

We have installed Elasticsearch, Wazuh manager, and Kibana on central management server. Wazuh is undoubtedly one of the best host intrusion detection system (HIDS), but that's not enough. We want our intrusion detection system (IDS) to capture network traffic and identify network-based intrusions. For that, we will use Suricata, which is a network intrusion detection system (NIDS) [68].

```
1 Install dependencies by running the following command:
```

```
2 # apt-get install libpcre3 libpcre3-dbg libpcre3-dev build-
    essential libpcap-dev   \
3 libnet1-dev libyaml-0-2 libyaml-dev pkg-config zlib1g zlib1g
    -dev \
4 libcap-ng-dev libcap-ng0 make libmagic-dev                \
5 libnss3-dev libgeoip-dev liblua5.1-dev libhiredis-dev
    libevent-dev \
6 python-yaml rustc cargo
7
8 Now, we will install Suricata from Ubuntu PPA. The OISF
    maintains a PPA suricata-stable for Ubuntu. To use it:
9 # sudo add-apt-repository ppa:oisf/suricata-stable
10 # sudo apt-get update
11 # sudo apt-get install suricata
12
13 Enable and start the Suricata service:
14 # systemctl daemon-reload
15 # systemctl enable suricata
16 # systemctl start suricata
```

Configure Suricata on CMS Server: Once the Suricata is installed and running, we first need to update the available ruleset. To do that, run the following command:

```
1 # sudo suricata-update
```

suricata-update is the official way to update and manage rules for Suricata. The command will download the ruleset into /var/lib/suricata/rules/ (Please note down the location, you can tweak or disable the available rule whenever required) [73].

Automatically updating rulesets

Updating the ruleset manually every time is not a practical solution. Still, at the same time, it is essential to keep the rulesets up to date so that Suricata can detect intrusions. To achieve the goal, we will take help from the cron job that executes suricata-update every Sunday at 00:00.

Open the terminal and enter the crontab -e command to do that. In the next windows, add the following line:

```
1 # 0 0 * * SUN suricata-update
```

One of the cool things about Suricata is that it offers using other rule-sets with suricata-update. To see the available ruleset, run the following command:

```
1 # sudo suricata-update update-sources
```

Then have a look at what is available:

```
# sudo suricata-update list-sources
```

Figure 7.4 shows the available ruleset with Suricata. One can enable any ruleset easily but want to keep an eye on the license part. The ruleset with license as commercial is not a free one and might need to put a secret code to enable the ruleset, which gets from the vendor after making payment. In contrast, one can enable a free ruleset with suricata-update.

For example, to enable the oisf/trafficid ruleset, you can run following command:

```
# sudo suricata-update enable-source oisf/trafficid
# sudo suricata-update
#  systemctl restart suricata
```

To see the currently active rulesets, use list-enabled-sources. Once you have identified and configured the correct ruleset, Suricata is ready to identify and protect the network from the external entity. Still, before that, we need to define our internal and external networks. One can do it by editing

```
Name: oisf/trafficid
  Vendor: OISF
  Summary: Suricata Traffic ID ruleset
  License: MIT
Name: ptresearch/attackdetection
  Vendor: Positive Technologies
  Summary: Positive Technologies Attack Detection Team ruleset
  License: Custom
Name: sslbl/ssl-fp-blacklist
  Vendor: Abuse.ch
  Summary: Abuse.ch SSL Blacklist
  License: Non-Commercial
Name: et/open
  Vendor: Proofpoint
  Summary: Emerging Threats Open Ruleset
  License: MIT
Name: scwx/security
  Vendor: Secureworks
  Summary: Secureworks suricata-security ruleset.
  License: Commercial
  Parameters: secret-code
  Subscription: https://www.secureworks.com/contact/ (Please reference CTU Countermeasures)
Name: scwx/malware
  Vendor: Secureworks
  Summary: Secureworks suricata-malware ruleset.
  License: Commercial
  Parameters: secret-code
  Subscription: https://www.secureworks.com/contact/ (Please reference CTU Countermeasures)
Name: et/pro
  Vendor: Proofpoint
  Summary: Emerging Threats Pro Ruleset
  License: Commercial
  Replaces: et/open
  Parameters: secret-code
  Subscription: https://www.proofpoint.com/us/threat-insight/et-pro-ruleset
```

Figure 7.4 Suricata available request

the Suricata configuration file, which is located at the following location /etc/suricata/suricata.yaml:

```
# nano /etc/suricata/suricata.yaml
```

Under the var section, you will need to change some critical variables such as HOME_NET and EXTERNAL_NET. By this time, Suricata is all set to flag detection. Still, we want to forward the traffic captured by Suricata to Wazuh manager for correlation and display the alerts on Kibana dashboard. By default, Suricata writes alerts to */var/log/suricata/eve.json* and the Wazuh agent does not monitor the location by default. To forward the Suricata logs to Wazuh manager, we have to add the following lines:

```
<localfile> configuration block in /var/ossec/etc/ossec.conf
    file:
<localfile>
<log_format>json</log_format>
<location>/var/log/suricata/eve.json</location>
</localfile>
```

Restart Wazuh manager service to apply the changes and that's it:

```
# systemctl restart wazuh-manager
```

7.3.7 Integration with IP repudiation feeds

IP reputation feeds are the reputation score system of an individual IP address, which businesses can use to detect risk or fraud associated with any IP. There are various open-source services for the IP reputation. For this project, we used IP feeds by FireHOL. FireHOL has several IP reputation feeds, but we used alienvault_reputation feeds. One can find more information from the source [26].

7.3.8 Configuring the CDB lists

To create the IP reputation database, we use a CDB list (constant database), a feature offered by Wazuh which can be used to create a list of any dynamic entity like users, file hashes, IPs, or domain names [79]. First, we must convert the blacklist format into the CDB list format. The format of the CBD list is key: value. We can take the help of a python script to do the work. The Python script will put the IP as the key, and the value will be empty. Note that the empty lines must be removed. Now, we download or pull the blacklist from FireHOL [42].

```
1 # sudo wget https://raw.githubusercontent.com/firehol/
    blocklist-ipsets/master/alienvault_reputation.ipset -O /
    var/ossec/etc/lists/alienvault_reputation.ipset
```

Now, we use the Python script provided by Wazuh community to convert it into a CDB list format:

```
1 # sudo wget https://wazuh.com/resources/iplist-to-cdblist.py
    -O /var/ossec/etc/lists/iplist-to-cdblist.py
2 # sudo chmod +x /var/ossec/etc/lists/iplist-to-cdblist.py
3 # sudo /var/ossec/etc/lists/iplist-to-cdblist.py /var/ossec/
    etc/lists/alienvault_reputation.ipset /var/ossec/etc/
    lists/blacklist-alienvault
4 # sudo rm -f /var/ossec/etc/lists/alienvault_reputation.
    ipset
5 The Python script will generate the blacklist-alienvault.
    Now, before using the generated list, we need to compile
    the list:
6 # sudo /var/ossec/bin/ossec-makelists
```

After compiling, the list the file blacklist-alienvault.cdb will be generated. Remember, here we are using the default directory for CDB lists: /var/os sec/lists and the newly created list must be defined in the /var/ossec/etc /ossec.conf file. To do that, we need to add below configuration block in the /var/ossec/etc/ossec.conf file:

```
1 <ruleset>
2 <list>/etc/lists/blacklist-alienvault</list>
3 </ruleset>
```

Using the CDB list in the rules:

As we have generated the list, we can start using our list in rules to get alerts. One can check the Custom Rules Section for that. Please restart the Wazuh manager service to apply the changes:

```
1 # systemctl restart wazuh-manager
```

7.4 Honeypots

7.4.1 WordPress honeypot deployment and configuration

Figure 7.5 shows the WordPress honeypot deployed.

```
1 Install Apache: Run following commands for apache server
    configuration \cite{digitalocean14}
2 sudo apt-get update
```

```
 3 sudo apt-get install apache2
 4 Apache2 -version
 5 Sudo service apache2 status
 6
 7 Install Database 'MySQL': Run following commands for MySQL
     server configuration
 8
 9 sudo apt install mysql-server
10 sudo mysql_secure_installation
11 Mysql -version
12 Sudo service mysql status
13
14
15 Install PHP': Run following commands for PHP on web server
     \cite{linuxize15}
16 sudo apt update
17 sudo apt install php libapache2-mod-php
18 php -version
19 sudo apt install phpmyadmin php-mbstring php-gettext
20 sudo phpenmod mbstring
21 sudo systemctl restart apache2
22
23 Install wordpress: Run following commands for Wordpress on
     web server  \cite{tecmint16}
24 wget -c http://wordpress.org/latest.tar.gz
25 tar -xzvf latest.tar.gz
26 sudo rsync -av wordpress/* /var/www/html/
27 sudo chown -R www-data:www-data /var/www/html/
28 $ sudo chmod -R 755 /var/www/html/
29 mysql -u root -p
30 mysql> CREATE DATABASE wp_myblog;
31 mysql> GRANT ALL PRIVILEGES ON wp_myblog.* TO '
     your_username_here'@'localhost' IDENTIFIED BY
     your_chosen_password_here';
32 mysql> FLUSH PRIVILEGES;
33 mysql> EXIT;
34 sudo mv wp-config-sample.php wp-config.php
35
36 MySQL settings - You can get this info from your web host.
37 /* The name of the database for WordPress */
38 define('DB_NAME', 'database_name_here');
39 /** MySQL database username */
40 define('DB_USER', 'username_here');
41 /** MySQL database password */
42 define('DB_PASSWORD', 'password_here');
43 /** MySQL hostname */
```

```
44 define('DB_HOST', 'localhost');
45 /** Database Charset to use in creating database tables.
46 */ define('DB_CHARSET', 'utf8');
47 /** The Database Collate type. Don't change this if in doubt
     .
48 */ define('DB_COLLATE', '');
49 $ sudo systemctl restart apache2.service
50 $ sudo systemctl restart mysql.service
```

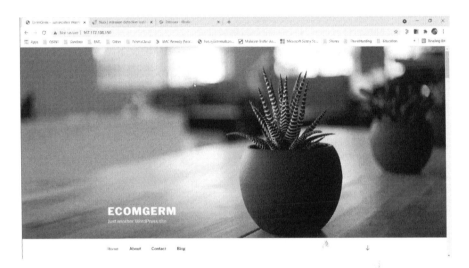

Figure 7.5 WordPress honeypot.

7.4.2 Mailoney honeypot deployment and configuration

Figure 7.6 shows the mailoney honeypot deployed.

```
1 git clone https://github.com/phin3has/mailoney \cite{
     mailoney17}
2 sudo apt update
3 apt install -y python3 python3-pip
4 mkdir -p /opt/mailoney
5 /usr/bin/pip3 install -r requirements.txt [WORKDIR /opt/
     mailoney]
6 mkdir -p /var/log/mailoney
7 touch /var/log/mailoney/commands.log
8 Python mailoney.py [-h] [-i <ip address>] [-p <port>] -s
     mailserver -t open_relay
```

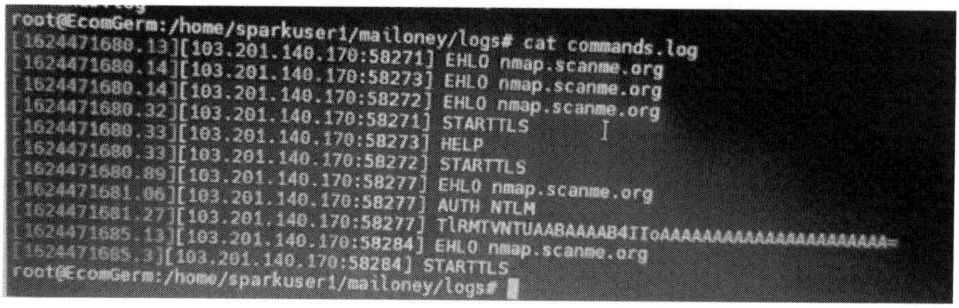

```
root@EcomGerm:/home/sparkuser1/mailoney/logs# cat commands.log
[1624471680.13][103.201.140.170:58271] EHLO nmap.scanme.org
[1624471680.14][103.201.140.170:58273] EHLO nmap.scanme.org
[1624471680.14][103.201.140.170:58272] EHLO nmap.scanme.org
[1624471680.32][103.201.140.170:58271] STARTTLS        I
[1624471680.33][103.201.140.170:58273] HELP
[1624471680.33][103.201.140.170:58272] STARTTLS
[1624471680.89][103.201.140.170:58277] EHLO nmap.scanme.org
[1624471681.06][103.201.140.170:58277] AUTH NTLM
[1624471681.27][103.201.140.170:58277] TlRMTVNTUAABAAAAB4IIoAAAAAAAAAAAAAAAAAAAAAA=
[1624471685.13][103.201.140.170:58284] EHLO nmap.scanme.org
[1624471685.3][103.201.140.170:58284] STARTTLS
root@EcomGerm:/home/sparkuser1/mailoney/logs# 
```

Figure 7.6   Mailoney honeypot.

7.4.3 Cowrie honeypot deployment and configuration

1 sudo apt-get update && sudo apt-get upgrade -y
2 sudo apt-get install git python-virtualenv libssl-dev libffi
    -dev build-essential libpython3-dev python3-minimal
    authbind virtualenv
3 Create a new user responsible for running the cowrie
    honeypot
4 $ sudo adduser cowrie
5 Now we need to delete the password for the user. This way
    nobody will be able to sign into the user, but we can
    still switch to the user with the su command:
6 $ sudo passwd -d cowrie
7 Switch user to the newly created cowrie user and change
    directory to the its home folder:
8 $ sudo su cowrie
9 $ cd
10 cloning the cowrie repository in the user.
11 git clone http://github.com/cowrie/cowrie
12 cd cowrie
13 Now we will setup a virtual environment so that we can
    install our Honeypot
14 virtualenv --python=python3 cowrie-env
15 source cowrie-env/bin/activate
16 Install some required python packages to the virtual
    environment:
17 (cowrie-env)$ pip install --upgrade pip
18 (cowrie-env)$ pip install --upgrade -r requirements.txt

There are ways an attacker may use to figure out that they are inside a Cowrie honeypot. The most notable sign is that there is only one directory under /home named richard. Another sign is that the server is using the hostname srv04 [19]. First, we need to change the hostname, which is done in

Cowrie's configuration file. Installing the configuration file of Cowrie inside the virtual environment is as follows:

```
1  (cowrie-env)$ cd /home/cowrie/cowrie/etc
2  (cowrie-env)$ cp cowrie.cfg.dist cowrie.cfg
3  (cowrie-env)$ nano cowrie.cfg
4  Remove the # from the following line and set hostname:
5  # hostname =
6  While we are in the config file, we can also update Cowrie
      to listen to port 22 directly. Remove the # from the
      following line and change 2222 to 22:
7  # listen_port = 22
8  Go down to the following line, and do the same - change 2222
      to 22:
9  listen_endpoints = tcp:22:interface=0.0.0.0
10 Save and exit.
11 Exit from the virtual environment and from the user cowrie
12 (cowrie-env)$ deactivate
13 $ exit
```

We have configured Cowrie to listen on port 22. However, this collides with two security issues:

- Only privileged users may listen on ports below 1024,
- A privileged user should not execute the Cowrie honeypot [27].

The solution to this dilemma is Authbind, which lets us override the first rule to let our unprivileged Cowrie user listen on port 22 directly:

```
1  $ sudo apt-get install authbind
2  $ sudo touch /etc/authbind/byport/22
3  $ sudo chown cowrie:cowrie /etc/authbind/byport/22
4  $ sudo chmod 770 /etc/authbind/byport/22
5  Change the execution file of Cowrie to reflect that it is
      going to use Authbind:
6  $ sudo nano /home/cowrie/cowrie/bin/cowrie
7  Change the following from ''no'' to ''yes'':
8  AUTHBIND_ENABLED=no
9  Switch back into the user cowrie again, because root should
      not start cowrie:
10 $ su cowrie
11 Start Cowrie and exit back to the original user:
12 $ /home/cowrie/cowrie/bin/cowrie start
13 $ exit
14 Some useful places to know about
15 Cowrie:
16 Logs: /home/cowrie/cowrie/log
```

```
17 Captured binaries: /home/cowrie/cowrie/dl
18 Playback files: /home/cowrie/cowrie/log/tty
```

7.4.4 Dionaea honeypot deployment and configuration

SSH into your honeypot and begin with the following commands to ensure that the distribution is up-to-date [24]:

```
1 $ sudo apt-get update
2 $ sudo apt-get dist-upgrade
3 $ sudo apt-get install inotify-tools
```

Make sure that the necessary tools are available in order to easily manage personal package archive (PPA) resources:

```
1 $ sudo apt-get install software-properties-common
2 git clone https://github.com/DinoTools/dionaea.git
3 cd Dionaea
4 sudo apt-get install build-essential cmake check cython3
       libcurl4-openssl-dev libemu-dev libev-dev libglib2.0-dev
       libloudmouth1-dev libnetfilter-queue-dev libnl-3-dev
       libpcap-dev libssl-dev libtool libudns-dev python3.8
       python3.8-dev python3-bson python3-yaml python3-boto3
       fonts-liberation -y
5 sudo apt-get install libemu-dev
6 mkdir build
7 cd build
8 cmake -DCMAKE_INSTALL_PREFIX:PATH=/opt/dionaea
9 make
10 sudo make install
11 The new honeypot can be found in the directory /opt/Dionaea
12 sudo /opt/dionaea/bin/dionaea start
13 Dionaea:
14 Logs: /opt/dionaea/var/dionaea
15 Captured binaries: /opt/dionaea/var/dionaea/binaries
16 Session transcripts: /opt/dionaea/var/dionaea/bistreams/YYYY
       -MM-DD
```

7.4.5 Deploying Wazuh agents on honeypot systems

The Wazuh agent is multi-platform and runs on the hosts that the user wants to monitor. It communicates with the Wazuh manager, sending data in near real-time through an encrypted and authenticated channel [78].

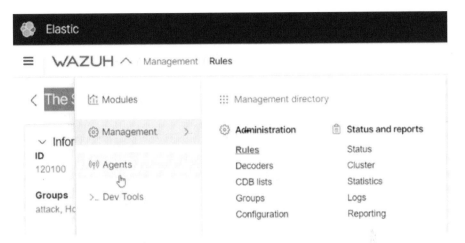

Figure 7.7   Wazuh Kibana dashboard agent tab.

Deploying Wazuh agents on Linux systems:

- Log in to Kibana console using wazuh_admin user,
- Click on Wazuh App and click on Agents tab in Wazuh App as shown in Figure 7.7,
- In the Agent panel, click on the deploy new agent as shown in Figure 7.8,
- Choose the operating system on which you want to install the agent,
- Type the Wazuh server address,
- Copy the command shown in Figure 7.9. Go to the endpoint on which you want to install the agent and run the command in terminal or command prompt depending on OS,
- Running the command will install the agent on the honeypot system, and it will get enrolled on the Wazuh App under agent tab.

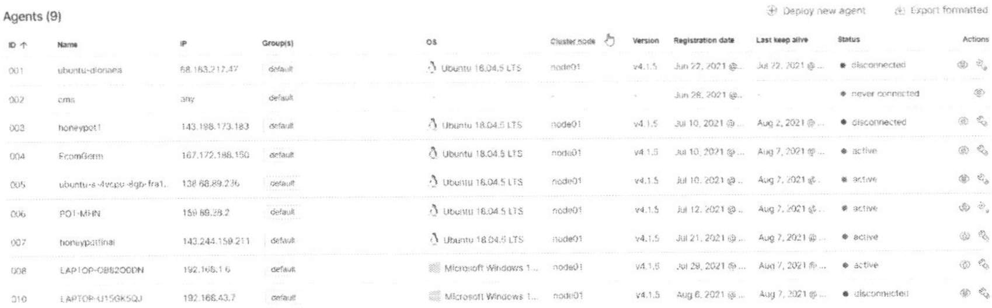

Figure 7.8   Deploy new agent.

Figure 7.9 New agent install and enroll process.

7.4.5.1 Configuration for logs forwarding from honeypot's server to Wazuh manager

To enable log forwarding, add the following configuration block in the ossec.conf file located at /var/ossec/etc/ on honeypot server. In the ossec.conf file look for $< ossec\_config >$ section and add the following configuration block there.

```
1  <localfile>
2    <local_format>apache</local_format>
3    <location>var/log/apache2/error.log </location>
4  </localfile>
5
6  <localfile>
7    <local_format>apache</local_format>
8    <location>var/log/apache2/access.log </location>
```

```
 9 </localfile>
10
11 <localfile>
12   <local_format>json</local_format>
13   <location>var/log/mysql/error.log </location>
14 </localfile>
15
16 <localfile>
17   <local_format>json</local_format>
18   <location>var/log/mysql/mysql.log </location>
19 </localfile>
20
21 <localfile>
22   <local_format>syslog</local_format>
23   <location>var/log/mailoney/commands.log </location>
24 </localfile>
25
26 <localfile>
27   <local_format>json</local_format>
28   <location>var/log/mysql/error.log </location>
29 </localfile>
```

7.4.6 Installing Suricata on honeypot server

Before installing Suricata, one should install Suricata's dependencies. One can install all the required dependencies by running the following command:

```
 1 apt-get install libpcre3 libpcre3-dbg libpcre3-dev build-
      essential libpcap-dev      \
 2 libnet1-dev libyaml-0-2 libyaml-dev pkg-config zlib1g zlib1g
      -dev \
 3 libcap-ng-dev libcap-ng0 make libmagic-dev              \
 4 libnss3-dev libgeoip-dev liblua5.1-dev libhiredis-dev
      libevent-dev \
 5 python-yaml rustc cargo
 6 Install Suricata from Ubuntu PPA
 7 For Ubuntu, the OISF maintains a PPA suricata-stable that
      always contains the latest stable release \cite{
      atlantic22}.
 8 To use it:
 9 sudo add-apt-repository ppa:oisf/suricata-stable
10 sudo apt-get update
11 sudo apt-get install suricata
12 Enable and start the Suricata service:
```

```
13 systemctl daemon-reload
14 systemctl enable suricata
15 systemctl start suricata
```

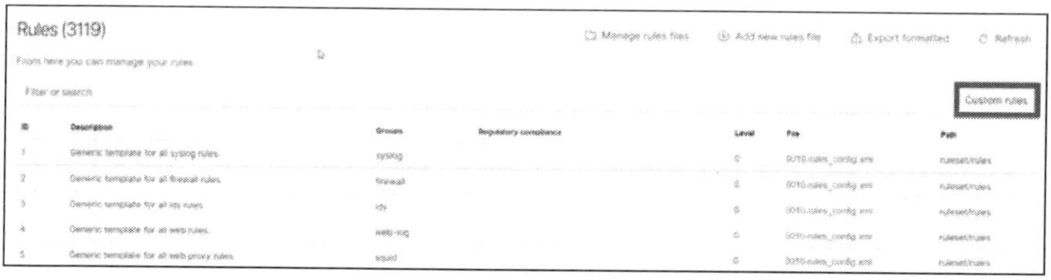

Figure 7.10   Custom rules.

7.4.7 Custom rules

In this section, we will learn to create custom rules on Wazuh manager or edit Wazuh's existing rules as per need or requirement. It can be achieved using the Wazuh Kibana plugin on the Kibana console.

Using Wazuh Kibana plugin interface

- Login to the Kibana console and open the Wazuh Kibana plugin interface.
- Then goes into the management tab and selects the rules. Then click on custom rule on the right side of the page as shown in Figure 7.10.
- Here, you have to click on add new rule file, and you can create a new rule with GUI. With a web interface, you will get an error message automatically if the XML syntax is not correct.
- Once we finish writing the rule, we have to click on Save file to confirm and then click on Restart now to restart the Wazuh manager service.

Adding custom rules for IP reputation feeds

We have created an IP_Reputation.xml rule file at /var/ossec/etc/rules/ IP_Reputation.xml.

following is the custom rule:

```
1
2 <group name="ipreputation,">
3
4  <rule id="120100" level="13"> <!-- Please note the level
      here, we have mentioned to push alerts from and above
      level 13 during Slack integration --!>
5    <if_group>web|attack|attacks|Honeypot|sshd|ids|suricata
      </if_group>
```

```
6    <list field="src_ip" lookup="address_match_key">etc/
     lists/blacklist-alienvault</list>
7    <description>The Source IP is in Alienvault black list
     .</description>
8      <mitre>
9        <id>T1110</id>
10       <id>T1037</id>
11     </mitre>
12   </rule>
13
14   <rule id="120102" level="13">
15     <if_group>web|attack|attacks|Honeypot|sshd|ids|suricata
     </if_group>
16     <list field="dst_ip" lookup="address_match_key">etc/
     lists/blacklist-alienvault</list>
17     <description>The Destination IP is in Alienvault black
     list.</description>
18       <mitre>
19         <id>T1110</id>
20         <id>T1037</id>
21       </mitre>
22   </rule>
23
24 </group>
```

Adding custom rules to detect Cowrie honeypot events. This is an example rule for Cowrie honeypot cowrie.login.failed events.

```
1
2  <group name="sshd,">
3
4    <rule id="120000" level="15">
5      <match>cowrie.login.failed</match>
6      <description>Cowrie SSH Login Failed Attempt</
     description>
7        <mitre>
8        <id>T1190</id>
9        <id>T1110</id>
10       </mitre>
11     <group>authentication_failed,invalid_login</group>
12   </rule>
13
14   <rule id="120001" level="15">
15     <match>cowrie.command.input</match>
16     <description>Cowrie SSH Exploit Detected</description>
17       <mitre>
18         <id>T1190</id>
```

```
19        <id>T1110</id>
20      </mitre>
21      <group>exploit_attempt ,Honeypot</group>
22    </rule>
23
24  </group>
```

Similarly, you can create custom rules for following Cowrie events:

- **cowrie.client.fingerprint** : If the attacker attemps to log in with an SSH public key this is logged here.
- **cowrie.login.success** : To detect successful authentication.
- **cowrie.client.size**: Width and height of the users terminal as communicated through the SSH protocol.
- **cowrie.session.file_upload**: File uploaded to Cowrie, generaly through SFTP or SCP or another way.
- **cowrie.session.connect**: New connection
- **cowrie.client.version**: SSH identification string
- **cowrie.client.kex** : SSH key exchange attributes
- **cowrie.session.closed** : Session closed
- **cowrie.log.closed**: TTY Log closed
- **cowrie.direct-tcpip.request** : Request for proxying via the honeypot
- **cowrie.direct-tcpip.data**: Data attempted to be sent through direct-tcpip forwarding

7.4.8 Centralized configuration

Centralized configuration is again a Wazuh feature using that we can manage and configure Wazuh agents remotely. The centralized configuration file agent.conf located at /var/ocssec/etc/shared/default on Wazuh manager.

We used the centralized configuration feature for log data collection from every honeypot and endpoints, security configuration assessment, file integrity monitoring and VirusTotal integration.

7.4.9 Log data collection

We have configured the log analysis engine to monitor Suricata logs files. To make it work, edit the file in Wazuh manager /var/ossec/etc/shared/default/agent.conf.

You can add the below block of configuration at the end of your agent.conf file and make it look like this:

```
1
2 <localfile>
3 <location>/var/log/suricata/eve.json</location>
4 <log_format>json</log_format>
5 </localfile>
```

Note that whenever you make any change in the agent.conf file, it is required to check for any configuration errors to avoid unintentional changes. You can confirm the configuration is valid by running verify-agent-conf present at /var/ossec/bin/verify-agent-conf on Wazuh manager.

```
1 # /var/ossec/bin/verify-agent-conf
```

Once the configuration changes validation has been completed, the Wazuh manager pushes the configuration files to all the agents. To do that, restart the Wazuh manager service.

```
1 # systemctl restart wazuh-manager
```

7.4.10 Security configuration assessment

The security configuration assessment (SCA) module is a set of predefined policies for system hardening and configuration guidelines. The policies are specific to the operating system the Wazuh agent has been installed and running.

The SCA predefined policies are based on CIS benchmarks. (add the URL as the reference here https://www.cisecurity.org/cis-benchmarks/)

To enable the SCA module, add the following block in Centralize Configuration file /var/ossec/etc/shared/default/agent.conf present on Wazuh manager:

```
1
2 <sca>
3     <enabled>yes</enabled>
4     <scan_on_start>yes</scan_on_start>
5     <interval>12h</interval>
6     <skip_nfs>yes</skip_nfs>
7 </sca>
```

To allow SCA policies to be pushed on any Wazuh agent remotely, we will have to configure it to accept the remote command and run the following command on all honeypot servers. This configuration change is mandatory.

```
1
2 # echo "sca.remote_commands=1" >> /var/ossec/etc/
       local_internal_options.conf
```

7.4.11 File integrity monitoring

The file integrity monitoring (FIM) module monitors selected files or directories and triggers alerts when any change is observed. By default, file integrity ,monitoring looks for selected directories and files depending on the host's operating system.

The FIM module is beneficial, and there are different use cases based on the configuration, but we will discuss the real-time monitoring we have used here.

Configuring real-time monitoring

We configured dionaea honeypot to capture the malicious files dropped by an attacker when he connects to dionaea honeypot. We used the real-time monitoring feature to watch for the directory where the malicious files were dropped. The real-time monitoring in the FIM module will trigger an alert as soon as it detects any new file. The real-time monitoring can be configured by real-time attribute.

Please note that the real-time attribute only works for directories and not with files. To enable the real-time monitoring for the dionaea binaries directory, add the below configuration block in the Centralized Configuration file /var/ossec/etc/shared/default/agent.conf present on Wazuh manager:

```
1
2 <syscheck>
3   <directories check_all="yes" realtime="yes">/opt/dionaea/
       var/dionaea/binaries</directories>
4 </syscheck>
```

7.4.12 VirusTotal integration

File integrity monitoring module watches for any modification or addition in a directory. This feature can be used to detect files with a malicious verdict by integrating it with the VirusTotal platform. File integrity monitoring permanently stores the hash of the file and triggers alerts when any new file is added or changes are made to the original file. After the VirusTotal integration is enabled, the FIM alert then makes an HTTP POST request to the VirusTotal using the VirusTotal API to look for the extracted hash in the VirusTotal database as shown in Figure 7.11. To enable the VirusTotal integration, add the following configuration block in /var/ossec/etc/ossec.conf file on your Wazuh manager.

```
1 <integration>
2   <name>virustotal</name>
```

```
3   <api_key>API_KEY</api_key> <!-- Replace with your
        VirusTotal API key -->
4   <group>syscheck</group>
5   <alert_format>json</alert_format>
6  </integration>
```

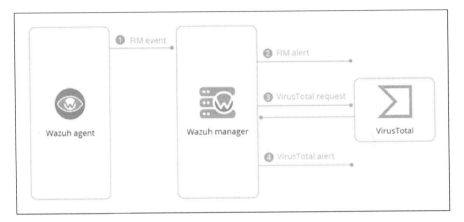

Figure 7.11   File integrity monitoring with VirusTotal integration.

If you do not have your VirusTotal API key, you can get your public or private API key by signing up to VirusTotal Community. (https://www.virustotal.com/gui/join-us).

After applying the configuration, restart the Wazuh manager service:

```
1  systemctl restart wazuh-manager
```

Now, to see the VirusTotal alerts in Kibana dashboard, enable the VirusTotal module in Wazuh App. To do that,

- Log in to Kibana and navigate to Wazuh App > Wazuh Settings > Modules
- Scroll down to Threat Detection and Response section
- Enable the VirusTotal module here.

Once enabled, we can see alerts in the VirusTotal dashboard.

Terms of Service

VirusTotal's Terms of Service specify the two ways the VirusTotal API may be used:

Public API

This method uses a free API with many of VirusTotal's functionalities. However, it has some significant limitations, such as: the request ratio limitation to no more than four requests per minute, and low priority access of requests done by this API for the VirusTotal engine. The VirusTotal documentation indicates that users who run a honey client, honeypot, or any other automation that provides resources to VirusTotal are rewarded with a higher request rate quota and special privileges when performing the calls to the API.

Private API

VirusTotal also provides a premium private API where the user's Terms of Service limit the request rate and the total number of queries allowed. Apart from that, it provides high priority access for requests and additional advantages.

It is essential to understand which configuration file takes precedence between ossec.conf and agent.conf when the central configuration is used. When the central configuration is utilized, the local and the shared configuration are merged. However, the ossec.conf file is read before the shared agent.conf, and the last configuration of any setting will overwrite the previous.

7.4.13 Slack

Slack is a great messaging tool widely used worldwide because of its capabilities like messaging in channels, forwarding notifications, and integrating third-party tools. Wazuh manager provides a webhook method to allow integration with Slack. This integration forwards the alerts generated on Wazuh manager to the Slack channel, which will trigger a notification on the mobile or laptop.

7.4.13.1 Integration with Slack

To start the integration process, start with the Wazuh's Integrator Daemon, which will allow Wazuh manager to connect to external APIs of Slack [71]:

Sending messages using Incoming Webhooks

Slack's Incoming Webhook is the simple way to post messages in Slack channel. Create a new Incoming Webhook to forward alerts to Slack channel.

- Navigate to this URL: https://api.slack.com/apps?new_app=1
- Login to Slack account if one sees You'll need to sign in to your Slack account to create an application. message on screen
- We will get a new window to create a Slack App.
- In the app name field, give your application a meaningful name.
- In the Development Slack Workspace field, select your Slack workspace.
- Click Create App button.

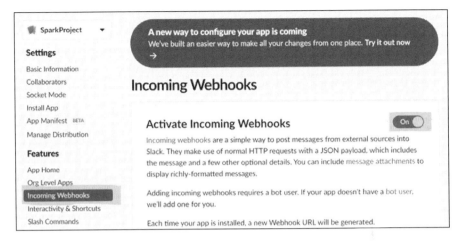

Figure 7.12 Slack Incoming Webhook setting.

Enable Incoming Webhooks: After creating the App, it will redirect to the settings page. On the settings page, select the Incoming Webhooks feature [70] as shown in Figure 7.12. Then, click the activate Incoming Webhooks toggle to switch it on.

Now that Incoming Webhooks are enabled, refresh the page, and some extra options will appear automatically. Scroll down and click on add new webhook to workspace button. We will redirect to the app settings page, and here we will get our webhook URL for the selected workspace section. The webhook URL should look something like this:

```
https://hooks.slack.com/services/XXXXXXXXXX/XXXXXXXXXX/
     XXXXXXXXXXXXXXXXXXXXXXXX
```

7.4.14 Configuration on Wazuh server

The integrations are configured on the Wazuh manager's /var/ossec/etc/ ossec.conf file. To configure an integration, add the following configuration block inside the $< ossec\_config >$ section:

```
<integration>
  <name>slack</name>
  <hook_url>https://hooks.slack.com/services/...</hook_url>
     <!-- Replace with your Slack hook URL -->
  <alert_format>json</alert_format>
```

```
5   <level>13</level> <!-- Please note that only alerts with
       the specified level or above are pushed to Slack.
       Allowed value is from 0 to 16 -- >
6   </integration>
```

After enabling the Daemon and configuring the integrations, restart the Wazuh manager service to apply the changes: systemctl restart wazuh-manager

7.5 Result

7.5.1 Geolocation of attacks

We have deployed our honeypot servers in different locations over the globe, like Germany, India, etc. and received multiple attacks on our honeypot service from various places where we found China is the most active and attacking country on the globe. Figure 7.13 shows the corresponding statistics.

7.5.2 Top usernames

In our study, we found attackers are attempting with the most common user's name during the brute force attack, so we should avoid these usernames in account creation action. Figure 7.14 shows the top usernames accessed.

Figure 7.13   Received attacks on honeypot servers over the globe.

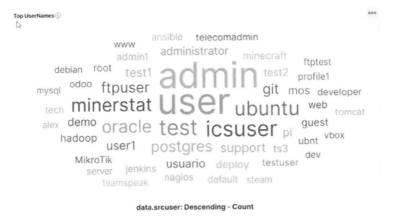

data.srcuser: Descending - Count

Figure 7.14 Top attempted usernames.

7.5.3 Top attacker countries

In our study, we observed that China is one of the most attacker countries globally. We received different types of multiple attacks on our honeypot servers. The United States is the second most active and attacker country. Figure 7.15 shows the details for the top countries that accessed our honeypots.

7.5.4 Top 10 attacker machine IPs

As we already saw in the previous result, China is the most active attacker country globally. All top 10 attacking machines belong to China, and we

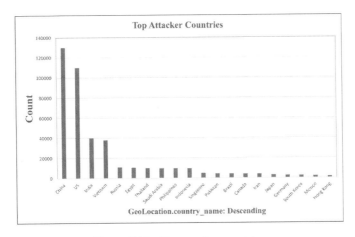

Figure 7.15 Top attacker countries.

Top 10 Attacker IPs

Source IP	Country	Attempt Count	Attempt Count percentages ▾
49.88.112.72	China	13,605	43.137%
49.88.112.71	China	5,064	16.056%
49.88.112.76	China	4,360	13.824%
222.187.239.109	China	1,745	5.533%
218.92.0.207	China	1,656	5.251%
221.181.185.223	China	1,495	4.74%
222.186.30.112	China	1,260	3.995%
221.181.185.151	China	1,195	3.789%
222.187.238.136	China	1,159	3.675%

Figure 7.16   Top attacker machines IPs.

received 43.137% attacks from 49.88.122.71 (Attacker IP) on our honeypot servers. Figure 7.16 shows the corresponding results.

7.5.5 Attacks for MySQL servers

WordPress honeypot server is running LAMP architecture, so we received almost all major SQL statements on our honeypot server where attackers try to get the data from the database. We observed most executable SQL is "INIT DB wordpress". Figure 7.17 shows the various SQL statements.

7.5.6 Top signatures-based attempts

Suricata inspects the live network traffic and detects signature-based attacks. We found port 443 is the most attempting port number by the attacker, where we received approx 0.13k attacks. Figure 7.18 shows the top signature-based attempts.

7.5.7 Top 10 attempted passwords

In our study, we found attackers are attempting with the most common user's password during the brute force attack, so we should avoid these passwords in account creation action and follow OWASP guidelines for better password selection. Figure 7.19 shows the top attempted password.

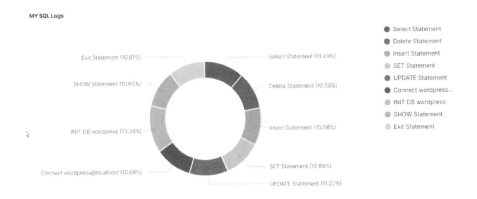

Figure 7.17   Top executed SQL statements.

signatureDetails ⓘ

data.alert.signature: Descending	data.src_ip: Descending	data.src_port: Descending	Count
SURICATA STREAM Packet with invalid ack	69.55.54.251	443	12,952
SURICATA STREAM Packet with invalid ack	69.55.54.251	53484	4
SURICATA STREAM Packet with invalid ack	103.157.22.49	1043	504
SURICATA STREAM Packet with invalid ack	103.157.22.49	1030	463
SURICATA STREAM Packet with invalid ack	103.157.22.49	1029	367
SURICATA STREAM Packet with invalid ack	103.157.22.49	1031	277

Figure 7.18   Top signatures-based attempts on honeypot and CMS servers.

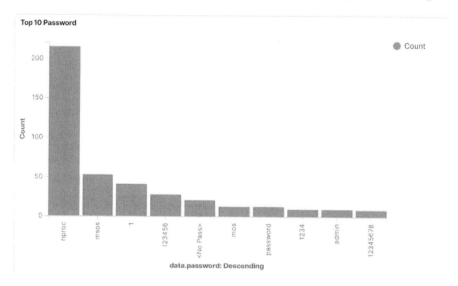

Figure 7.19   Top attempted passwords.

7.6 Conclusion and Future Work

Our study explained deception technologies and implemented low, medium, and high interaction honeypots on clouds. It developed HIDS and NIDS on top of it using open-source tools like Wazuh, Suricata, Open distro, etc. Our goal is to understand the attacker's intention and how they are working to lure intruders toward the production system. We discovered their security flaws to help small-or medium-scale industries at a feasible cost. Real-time threat analytics will help organizations monitor real-time attacks and protect their system according to the business requirement of the product. We have analyzed cyber threat intelligence using honeypot logs collected from DigitalOcean clouds. The data is analyzed using the Open Distro stack for log data visualization. It is worth noting that Open Distro uses Elasticsearch, which helps to identify various types of cyber incident events through Wazuh. It has become apparent in the present time; attackers are constantly targeting honeypots. Most of the attacks are similar as attackers attempt to gain full access to the system. This study into honeypot logs for cyber threat intelligence is valuable as it can be used to identify and mitigate future cyberattacks related to your business logic. The main advantage of using honeypot logs for threat intelligence is that there is no impact on the production system. This kind of analysis could always help build future IDS and IPS for production. The implementation of deception technology and intrusion detection system gives us ability to go beyond the traditional signature-based approach for intrusion detection.

We have used Suricata to capture and analyze network traffic in this work. We know about the Suricata's network detection capabilities, but we want to extend the scope to prevent the intrusions at the perimeter level using Suricata's prevention system. Because Suricata can work as intrusion prevention system and help prevent the environment from known network attacks and anomalies. Similarly, the Wazuh can block the attacks using the active response feature present in Wazuh. The active response allows the Wazuh agent to run commands on the end system, and we can use this feature to write a custom script to block IP or perform any other action. We planned to configure Suricata to block network attacks and Wazuh to block host anomalies as and when detected.

In this work, we discuss different honeypots and managing them is a challenge in the real world. We experience that honeypots are getting down due to a rapid increase in traffic or exploit attempts. We missed the attacker footprints until the time we brought up the honeypot manually. To deal with such unavoidable circumstances, we have planned the honeypot orchestration. We want to automate the deployment and management of honeypot using the Kubernetes orchestration framework.

Last but not least, we plan to improve the detection capabilities of our work by adding machine learning-based detection. A signature-based intrusion detection system comes with the risk of missing the attack or anomaly for which we do not have a signature in the database. However, with ML-based detection, we can analyze the traffic's patterns or behavior and trigger detection based on the confidence value about any uncertain attack. Combining the deception technology and the ML-based detection will help capture and analyze zero-day attacks and vulnerabilities. We can train the ML model again, and this will be a continuing process till we achieve accuracy.

8

Attack Vector Analysis with a New Benchmark

Ashish Ranjan Yadav and Rohit Negi

C3i Center, Indian Institute of Technology, Kanpur, India
E-mail: aryadav@cse.iitk.ac.in; rohit@cse.iitk.ac.in

Abstract

Gone are the days when Linux users assumed Windows was the main target of hackers. Apart from Windows-based systems, Linux-based devices are becoming a more valuable target. Attackers are working on new ways to compromise Linux-based distributions on supercomputers, cloud servers, and various IoT devices. The crypto mining virus Lemon Duck recently compromised Linux systems using SSH brute-force assaults. A port scanning module of the malware searches for internet-connected Linux computers listening on the 22 TCP ports used for SSH remote login. It has become critically important to keep a close check on the application, network devices, firewalls, and configuration changes in the system. Any minor configuration changes that do not adhere to security norms can open the system to attacks. To improve the cybersecurity posture of the current system, we have implemented system hardening as a technique to reduce the system's attack surface. With the increasing size of the organization and the number of users, it has become difficult for a system admin to monitor each system manually for the security audit. This work will enable the system admin to monitor the system configuration and security audit in near real-time, thus enhancing the current cybersecurity posture.

8.1 Introduction

Servers are always present at the most prominent IT security risk. They store and process every business information, from business operation data to critical user information to financial records. Most of the application and database security is dependent on how secure the server is. Hence IT security teams need to implement sophisticated methods that build strong security postures to stay ahead of the threat actors. It is understandable why most businesses prioritize server security over other aspects of their IT infrastructure. Most of the bounty that threat actors intend to steal is stored on servers.

Most of the servers are protected by firewalls, and IDS is configured to enhance the security posture further. Servers are sometimes left unsecured, leaving them vulnerable to massive attacks. In November, CISOMAG published a report on how security researchers uncovered an open Elasticsearch server with 1.2 billion unique data records. The server housed more than four gigabytes of data without password security or authentication.

PgMiner botnet attacks in December 2020 happened by launching brute-force assaults against Internet-accessible PostgreSQL databases, according to researchers at Palo Alto Networks' Unit 42. The botnet selects a public network range at random and then searches all IP addresses for systems with the PostgreSQL port (port 5432) accessible online. If PgMiner detects an active PostgreSQL server, the botnet switches from scanning to brute-force mode, where it shuffles through a huge list of passwords to guess the credentials for the default PostgreSQL account as, "postgres." Suppose the owner of a PostgreSQL database forgets to disable or change this user's credentials. In that case, the hackers gain access to the database and utilize the PostgreSQL COPY from PROGRAM functionality to escalate their access from the database app to the underlying server, allowing them to take control of the entire OS.

A privilege escalation vulnerability was reported in the default Ubuntu Linux installed in January 2019. It was caused by a flaw in the snapd API, which is a standard service. Any local user could exploit this vulnerability to get immediate root access to the machine.

Patients' personal information was compromised due to a cyberattack at Mississippi's Coastal Family Health Center (CFHC). Patients' confidential medical information from Northwestern Memorial HealthCare (NMHC) providers may have been exposed due to a data breach at a third-party provider. Unknown individuals got unauthorised access to a database controlled by Elekta, a cloud-based platform that manages Illinois' legally mandated cancer reporting. A third-party cloud provider was attacked, exposing 190,000 patients' data at US healthcare organizations. In May 2020, criminals took control of Blackbaud's servers and encrypted some of the company's data in a ransomware assault.

A server admin is generally responsible for hosting a new application and updating the system with the latest security patches. Sometimes installing new packages and updating the system might open some backdoor to exploit the system and make the system vulnerable and prone to attack. Most places focus on securing the network and firewall but not on the application and packages. We focus on securing the system with vulnerable packages and updates with our work. We have implemented system hardening on the system and given near real-time feedback on system hardening scores with the help of the lynis tool.

Before installing any package or updates on the server, the packages are first installed on a sandboxed hardened Linux system. Based on the score returned user can proceed with the installation on a real system or discard the installation.

Configuring an OS securely, creating rules and policies, updating it to control the system securely, and deleting superfluous apps and services are all examples of hardening the OS. It reduces a computer OS's vulnerability to threats and mitigates potential dangers.

System hardening is changing system configuration and state to reduce system vulnerability. The main goal of hardening is to minimize the system's attack surface. It comprises tools and a set of practices to minimize the threat vector of the system. Hardening a system does not ensure that the system will not be attacked or vulnerability-proof. It reduces the system's risk vector to a greater extent to protect it from well-known attacks. No system is entirely secure, and there is always a chance for improvement. With our work, we are not claiming our system to be unbreachable.

Our work is based on the Linux system. Linux system comprises of Linux kernel and operating system. Linux system is already considered safe as most Linux distribution comes up with various security-related tools, and many security features are built into the kernel. Linux systems are infinitely configurable, and even a small configuration change can significantly impact security. Hence we should go very carefully with the changes we are making to the system as a lack of understanding can lead to unintentional exposure.

We have used the lynis tool to measure the system hardening index for our work. Lynis is a security audit tool for operating systems like Linux, macOS, or Unix-based systems. It scans the system for system hardening and compliance testing.

The idea is to create a Ubuntu sandbox using oracle VMBOX. System hardening steps are performed on the ubuntu VM to get a score of 96. Whenever a user tries to install some package on the host machine, the package will be first installed inside the sandbox. Sandbox will install the package and will check the lynis score post-installation. A decrease in lynis score means the package is making changes to the system that compromises the system's security. After returning the lynis score, it is up to the user to proceed with the installation. If the installation does not affect the lynis score, the package is not affecting the system hardening and is safe to use on the host machine. Hence, this work reduces the host system's attack surface by first installing the package on the sandbox and checking the score rather than directly installing the package on the host.

A similar idea is also imposed while purging or removing packages. The purge/remove operation will be first performed inside the sandboxed environment. Based on the returned lynis score, the action will be taken in a real environment. There might be the case that removing or purging any package will reduce the hardening index and make the system vulnerable to attacks. Hence, verifying the effects of purging or removing any package on the system is crucial. Figure 8.1 gives the overall idea of our work.

Our objective is to improve the current cybersecurity posture of the system by hardening the OS and giving time to time evaluation and validation of the hardening index. It helps identify vulnerabilities that arise by installing any new package to the system. Lynis's score returns the host post-installation of a package on the VM. It ensures real-time hardening resistance of the system. OS hardening reduces the attack surface to a greater extent, protecting the system from most known attacks and potential attackers. There is a reduction in the hardening index if there are any old or upgradable packages available

Figure 8.1   Overview of proposed work.

for the system. It ensures updated and secure infrastructure. System hardening is performed as per the suggestion from the CIS benchmark and additional information provided by the lynis. Hence, creating a new benchmark on top of the existing CIS benchmark.

8.2 Background and Related Work

System hardening is the set of practices through which attack surface and attack vectors can be reduced to enhance the security of servers and computer systems. System hardening involves closing system loopholes frequently used by the attackers to access the system and access sensitive user information.

System hardening involves deleting or disabling useless applications, protocols, file systems, ports, permissions, and other features that might put the system at risk of attack. Potential attackers will not be able to obtain access to the system.

Implementing system hardening reduces the potential doorways to the system an attacker might use to gain access to the system and exploit the system. System hardening involves securing computer applications and operating systems, firmware, databases, networks, and other system features that attackers might use to exploit the system. System hardening can be attained in various ways:

- Application hardening
- Operating System hardening
- Server hardening
- Database hardening
- Network hardening

The attack surface is all possible weak spots and backdoors in the system or network that attackers can exploit to compromise or access critical information. These vulnerabilities can arise due to several reasons, some of which are as follows:

- OEM.
- Unpatched or outdated software and firmware vulnerabilities.
- Default and hard-coded passwords.
- Users install password in plain text.
- Loosely configured networking devices like routers, switches along with unnecessary or unused ports and services.
- Unencrypted or weakly encrypted network traffics.
- Absence of privilege access control.

8.2.1 Application hardening

Application hardening involves enhancing the security of the server's applications like web browsers, text editors, and other applications. It consists in updating the application to the latest version or changing configurations, or modifying the source code related to the application. A few examples of application hardening are:

- OEM.
- Keeping application patched and updated.
- Using firewalls.
- Using encryption methods.
- Use of intrusion detection system or intrusion prevention system.
- Using antivirus, malware, and spyware protection application.
- Changing the configuration or attributes related to the application to enhance its security.

8.2.2 Operating system hardening

OS hardening involves enhancing the security of the server's operating system. It can be done by patching, updating the system to the latest version, and installing service packs. Make sure packages installed are from trusted repositories. Unlike application hardening's focus on securing standard and third-party applications, OS hardening protects the base software that grants programs access to specific tasks on your system. Examples of OS hardening are as follows:

- OEM.
- Partitioning the file system properly. For example in Ubuntu make sure /home /var /boot /tmp are in separate partition.
- Remove unnecessary drivers.
- Remove unused file system.
- Limit and authenticate system access permission.
- Enable secure boot.
- Encrypt file system.

8.2.3 Server hardening

Server hardening is a method of securing a server's data, ports, components, operations, and privileges by implementing sophisticated security mechanisms at the hardware, firmware, and software layers.

Some of the general server security measures are as follows:

- OEM.
- Server's operating system must be updated and patched regularly.
- Removing unused or unsafe third-party software. Update all the critical third-party software.
- Developing strong password policies for users and using stronger and more complicated passwords.
- Failed login attempts should be recorded and blocked.
- USB ports are disabled when the computer boots up.
- Putting in place multi-factor authentication.
- Advanced cybersecurity suites suited to the operating system, firmware resilience technologies, memory encryption, antivirus and firewall protection, and advanced cybersecurity suites adapted to the operating system are some of the methods used.

8.2.4 Database hardening

Database hardening refers to safeguarding both the contents of a digital database and the database management system (DBMS), which is the database application used by users to analyze and store data. Database hardening mainly involves three processes:

- OEM.
- Controlling and limiting user access and privileges.
- Useless or less frequently used database functions and services should be disabled.
- All the database resources must be encrypted.

Few of the database hardening techniques involve:

- OEM.
- Administrators' and administrative privileges and functions are restricted.
- Encrypting database data in transit and at rest.
- Following an RBAC (role-based access control) strategy.
- Patching and updating database software, or the DBMS, regularly.
- Disabling unnecessary database services and functions.
- Suspicious login activity to the database should be locked.
- Enforcing more difficult and strong database passwords.

8.2.5 Network hardening

Network hardening involves enhancing the security of system communication infrastructure to prevent unauthorized access from outside. It is one of the critical steps as most of the attacks are made from outside the network. Network hardening involves:

- Configuring and securing network firewalls.
- Regularly auditing the network rules.
- Securing remote access points and users.
- Blocking unused ports.
- Disabling and removing unnecessary network protocols and services.
- Implementing access lists.
- Encrypting network traffic.

Network hardening also involves establishing intrusion prevention or intrusion detection systems. These applications automatically monitor and report any suspicious activities in a given network which helps administrators prevent unauthorized access. Using the above methods reduces the network's overall attack surface and enhances its resistance to network-based attacks.

For system hardening, there are several industry standards and guidelines. Organizations like the Computer Information Security (CIS) Center for Internet Security, the National Institute of Standards and Technology (NIST), and Microsoft released best practices for hardening the system.

CIS Benchmark releases several configuration guidelines for web servers, different OS (like Ubuntu, Windows, Mac), cloud platforms, networking devices, and many more to reduce the system's attack surface and protect the system from several attacks.

The CIS Benchmarks are freely released in PDF format to promote their use and adoption as user-generated, de facto standards worldwide. Government, business, industry, and academia have produced and approved the CIS Benchmarks as the sole consensus-based, best-practice security configuration guides.

8.3 Threat Vector and Attack Surface

An attack is an act by which a threat actor gets access to an information system's assets. The path used to attack a system is called an attack vector. There are mainly three types of attack vectors and threats:

- **Network threats:** Refers to the threat against a network of the organization.

- **Host threats:** Threats against the host, including hardware and operating system.
- **Application threats:** Threats against the system programs.

8.3.1 Attack surface

The attack surface is the point in the system or network exposed to attack. Attack surfaces are the potential point of entry to the system for an attacker. Attack surface does not always mean digital; it can be physical. Attack surface analysis helps understand the risk areas in the system and makes security specialists aware of part of the system open to attacks to find ways of minimizing it. Organizations should focus on reducing attack surfaces to as minimum as possible. Figure 8.2 shows the attack surface.

The best approach to secure a system is to identify the high-risk areas. The focus should be on the remote access points like interfaces with the external systems and the Internet, i.e., the issues where the system allows access to the public. The next step should be to put some measures to safeguard the high-risk areas like network firewalls, application firewalls, and intrusion detection systems to protect the system. These will not make the system attack-proof. However, it will reduce the system's attack surface to a greater extent and

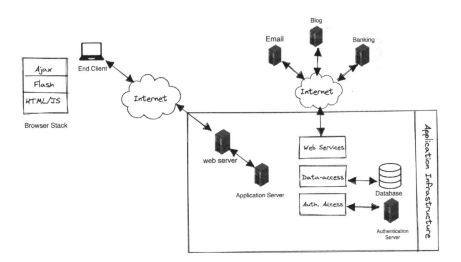

Figure 8.2   Attack surface.

protect the system from well-known attacks and, in some cases, also from zero-day attacks.

8.3.2 Attack vector

An adversary's method for breaching or infiltrating a network or system. Hackers can use attack vectors to take advantage of system flaws, including the human factor. Figure 8.3 shows the common attack vectors. Some of the common attack vectors are explained as follows:

Compromised credentials: The most common entry point for an attacker is through username and password. It is one of the most common forms of attack. When unauthorized entities gain access to user credentials such as username and password, this is referred to as compromised credentials. Such attacks are often brutal to detect as attackers impersonate real users with actual usernames and passwords. Compromised credentials result from irresponsible behaviour from users and admins. Most people still use credentials that are very easy to guess, and passwords are still stored in plain text format or inferior encryption methods prone to brute force attacks. Privilege access credentials that grant admin access to the system and device are considered most dangerous than consumer credentials. Such attacks can be avoided by enforcing strict password policies and frequently changing passwords. Password should be stored with robust encryption methods. The system should be frequently audited for any irregular behaviour.

Malicious insider: Malicious insiders are users within the organization who exploits the weakness in company infrastructure and organization

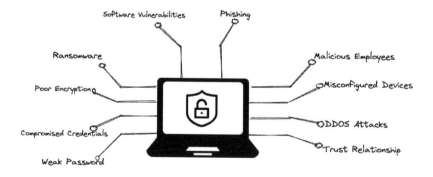

Figure 8.3   Common attack vectors.

information. Such users can cause significant damage to the system as they have privileged access to sensitive data and networks. It can be avoided by monitoring every user and device's data and network access inside the organization.

Encryption: Encryption converts plain data or text into a format that a person can only access with a key. Encrypted data is also known as ciphertext. Encryption is the need of age as all the data are being transmitted through network channels that attackers can easily access. Plain text data can be read and accessed by any user or attacker, putting sensitive information at risk. Poor or old encryption methods should also be avoided, as brute force methods can be used to break the encryption. Hence, robust encryption methods should be enforced at every data processing stage, especially while transmitting data over the network and data at rest in the case of servers and databases.

Misconfiguration: Misconfiguration is referred to as an error in system configuration. Hidden weaknesses or loopholes aroused in the system due to configuration or setup mistakes/errors can easily be detected by hackers to obtain more information about the system. Misconfigured systems or applications are considered easy access points for an attacker. Configure the system and network as per the recommended security norms. Configure application and device settings to enhance the security of the system.

Ransomware: Ransomware is a cyberattack where users are denied access to their system or data unless a ransom amount is paid to the attacker. After paying the ransom amount, users are provided with instructions to obtain the decryption key. The demanded amount depends upon how critical the information or system is. It varies from a few hundred dollars to thousands of dollars, and the amount is generally paid in the form of bitcoins. To protect the system from ransomware attacks, make sure the system is updated with the latest version of the software. Permissions and privileges to the applications should be closely monitored to precisely what it does.

Software Vulnerabilities: Malicious software is often granted full permission during installation. Software from official as well as unofficial app stores can have vulnerabilities. It is the user's responsibility to check the sanity of software before using it in the host or production environment. Users should be aware of changes made by the app in the system.

8.3.3 Hardening steps

Our work is based on Ubuntu 18.04 desktop as well as Server. For virtualization, we have used Oracle VMBOX, which is a virtual machine from Oracle.

Oracle VMBOX is installed with Ubuntu 18.04. Hardening is performed inside the Vbox, and the hardening index of the virtual machine is calculated by using the lynis tool. The hardening step starts as early as during the operating system installation on the system. There are various steps of hardening which are as follows:

8.3.4 During installation

The first step toward hardening starts with the partitioning of the file system during the installation of the OS. Make sure separate partition exists for **/home**, **/tmp** and **/var**. It is an important step, especially in a multi-user environment where the unprivileged user will have read/write access on partitions. It becomes easier to specify mount options separately for each partition with separate partitions, which further enhances the system's security. Generally, users will have two partitions, one for root **/** and another one is **/home**. This partition system will have a greater risk of resource exhaustion as /var partition could easily be filled by some misbehaved application which can further fill up the / partition and lead to a kernel crash. Similarly /tmp partition contains temporary data and might fill the / partition if /tmp is not kept on a separate partition. Hence, keeping /var and /tmp on separate partitions limit the amount of temporary data stored to prevent another important partitioning from filling up and causing system crashes and data loss.

8.3.5 Updates and upgrades

If the partition rules are followed as specified in the previous section, the lynis score on fresh installation is 58. There will be a few updates and upgrades available on a fresh installation. The first step post-installation of OS is to update and upgrade the packages present on the system. This step increases the lynis score to 63. Old packages are vulnerable to attacks; hence, keeping the packages updated to the latest version is crucial. It is considered the most common and simple step toward securing the system. Softwares regularly release updates or patches to include some additional features or rectify security flaws. Updates must be performed first before the upgrade, making apt aware of new versions of packages. Run the following command to update and upgrade all the packages:

```
1 apt update
2 apt upgrade
```

8.3.6 File systems

Linux supports several uncommon filesystem types. Disabling unused filesystems can help in reducing the attack surface of the system. Utmost care should be taken while using an uncommon filesystem as it may compromise the security and functionality of the system. Some unused filesystems in Linux are udf, hfs, hfsplus, freevfxs, jffs2, cramfs, etc. To disable uncommon filesystem on Ubuntu create or edit file under directory **/etc/modprobe.d/** with .conf extention. For example *uncommon.conf*. Run following commands in terminal:

```
1 echo "install cramfs/bin/true">>/etc/modprobe.d/uncommon.
    conf
2 echo "install udf/bin/true">>/etc/modprobe.d/uncommon.conf
3 echo "install hfs/bin/true">>/etc/modprobe.d/uncommon.conf
4 echo "install hsfplus/bin/true">>/etc/modprobe.d/uncommon.
    conf
5 echo "install jffs2/bin/true">>/etc/modprobe.d/uncommon.conf
6 echo "install freevfxs/bin/true">>/etc/modprobe.d/uncommon.
    conf
```

Mount options: To further reduce the attack surface of filesystem , strict mount options needs to be set on each partition. By default Linux will have *defaults* mount option set to all partition.The *defaults* is equivalent to rw, suid, dev, exec, auto, nouser, async(no ACL support). *noexec* mount option will prevent any binaries from executing on that partition. *nosuid* mount option will ignore setuid and setguid bit i.e., the partition cannot contain suid files. *nodev* mount option prevents partition from containing any device files.

To change the mount options of */tmp* partition, in the fourth field of /tmp partition inside */etc/ftsab* file, enter **nodev, nosuid, noexec**. Similarly set mount options **nodev, nosuid** for */home* partition. Mount options for */var* partition is set as **nodev, nosuid**. As separate partitions exist for */home, /var, /tmp* there will be entry present inside */etc/fstab*, mount options can be changed adding entries to fourth field.

/dev/shm functions as shared memory implementation on Unix systems which is used for sharing data between programs. It is used for inter-process communication purposes. If permitted, one program will create a memory portion, and another program can access the memory portion. shm is also known as a temporary file storage filesystem (tmpfs). Instead of persistent storage, it uses virtual storage. Make the following changes in */etc/fstab* file:

```
1 tmpfs /dev/shm tmpfs defaults, noexec, nodev, nosuid 0 0
```

Additional filesystem configuration to further enhance the security of system in addition to CIS benchmark involves setting mount options for */run,/dev,/proc,/var/log,/var/log/audit*. Figure 8.4 shows the few mount options in */etc/fstab* file. Edit */etc/fstab* file and add following line:

```
1  tmpfs /run tmpfs defaults, noexec, nodev, nosuid 0 0
2  tmpfs /dev tmpfs defaults, noexec, nosuid 0 0
3  tmpfs /pro tmpfs defaults, noexec, nodev, nosuid 0 0
4  tmpfs /run tmpfs defaults, noexec, nodev, nosuid 0 0
```

Figure 8.5 shows side by side comparison of lynis output on fresh installation and after adding mount options to partitions.

```
#
# <file system> <mount point>   <type>  <options>       <dump>  <pass>
# / was on /dev/sda1 during installation
UUID=3284ce66-8498-4501-9af7-a8fb8d4f6ea8 /              ext4    defaults 0      1
# /boot was on /dev/sda6 during installation
UUID=81d66874-140b-4e47-be72-db0527d68a04 /boot          ext4    nodev,noexec,nosuid      0      2
# /home was on /dev/sda7 during installation
UUID=b634117b-37db-431c-9bc0-0dffaedd2bfd /home          ext4    nodev,nosuid     0      2
# /tmp was on /dev/sda9 during installation
UUID=3fdcfcd9-5368-4e79-b48b-199b4e6245d9 /tmp           ext4    nodev,noexec,nosuid      0      2
# /var was on /dev/sda8 during installation
UUID=52a366d4-1289-428a-9458-278d2673004c /var           ext4    nodev,nosuid     0      2
# swap was on /dev/sda5 during installation
UUID=82160c99-79ba-4a71-8a0c-f1eec137f77e none           swap    sw       0      0

tmpfs                                      /dev/shm       tmpfs   defaults,nosuid,nodev,noexec 0 0
tmpfs                                      /run           tmpfs   defaults,nosuid,nodev,noexec 0 0
tmpfs                                      /dev           tmpfs   defaults,nosuid,nodev,noexec 0 0
tmpfs                                      /proc          tmpfs   defaults,nodev,nosuid,noexec,hidepid=1 0 0
tmpfs                                      /var/log       tmpfs   defaults,nodev,nosuid,noexec 0 0
tmpfs                                      /var/log/audit tmpfs   defaults,nodev,nosuid,noexec 0 0
/tmp /var/tmp none rw,noexec,nosuid,nodev,bind 0 0
```

Figure 8.4   Mount options in /etc/fstab file.

(a) Before hardening

(b) After hardening

Figure 8.5   Lynis output before and after defining mount points.

8.3.7 Users, groups and authentication

PAM (Pluggable authentication modules) implements authentication on a UNIX system. It is a collection of shared libraries that are used to authenticate a user to an application dynamically. PAM should be properly configured to enhance system authentication. Several attributes are present to strengthen the authentication. */etc/security/pwquality.conf* file have the following options :

- *minlen :* minimum password length.
- *minclass :* minimum number of types of character (lowercase, uppercase, digit, etc)
- *dcredit – n :* maximum allowed number of digit. If value is less than *n*, each digit will be counted toward *minlen*. Negative value of *n* means minimum number of digits required for password.
- *ucredit – n :* maximum allowed number of uppercase character for password. If value is less than *n*, each character will be counted toward *minlen*. Negative value of *n* means minimum number of uppercase character required for password.
- *lcredit – n :* maximum allowed number of lowercase character for password. If value is less than *n*, each character will be counted toward *minlen*. Negative value of *n* means minimum number of lowercase character required for password.
- *ocredit – n :* maximum allowed number of other character for password. If value is less than *n*, each character will be counted toward *minlen*. Negative value of *n* means minimum number of lowercase character required for password.

pam_pwquality can be installed by using following command :

```
apt install libpam-pwquality
```

Shadow password suite parameters: Many password parameters are configured in PAM , some password configuration are done in */etc/login.defs* file. *PASS_MIN_DAYS* parameter restricts users from changing their password within the mentioned number of days. This value should be at least 1. *PASS_MAX_DAYS* defines the age of the password. Limiting the password age will protect the system from brute-force attacks, reducing the attacker's window of opportunity. It is recommended that the value should not be greater than 365. *PASS_WARN_AGE* parameter notifies the user with a warning about the expiration of a password in a defined number of days. The value should

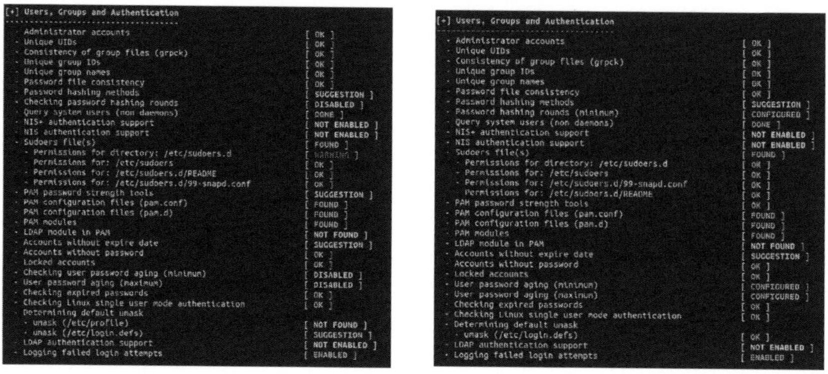

(a) Before hardening   (b) After hardening

Figure 8.6 Lynis output before and after user, group, and authentication.

be set to at least 7 days. The default value of *umask* is set to be 027 or more restrictive. Inside file */etc/profile*, set the value of *umask* to 027 .

```
1  PASS\_MIN\_DAYS 7
2  PASS\_MAX\_DAYS 365
3  PASS\_WARN\_AGE 7
4  UMASK 027
```

Inactive user accounts should be locked to prevent unauthorized access to the idle system. As per security norms, it is suggested that accounts inactive for more than 30 days should be blocked or disabled. To set the parameter run the following command:

```
1  useradd -D -f 30
```

Figure 8.6 shows the output of Lynis before and after user, group, and authentication.

8.3.8 Warning banners

Contents from */etc/issue* and */etc/issue.net* are displayed as welcome message when a user tries to remote login to the system. /etc/issue.net content is shown when a user tries to connect from the network. Message from /etc/issue is displayed when a local user tries to connect, and network user (if issue.net is not configured), **issue.net** displays message before the password prompt, i.e., before actually entering into the system. The **motd** content is displayed after the user has logged into the system. */etc/issue,/etc/issue.net* and */etc/motd* file contains OS version information by default, which is considered vulnerable

```
####################################################################################
#                    ALERT! You are entering into a secured area!               #
# Your IP,Login Time,Username has been noted and has been sent to the server administrator!#
# This service is restricted to authorized users only.All activities on this system are  #
# logged.Unauthorized access will be fully investigated and reported to the appropriate  #
#                          law enforcement agencies.                            #
####################################################################################
```

Figure 8.7   Content of */etc/issue./etc/issue.net.*

(a) Before hardening

(b) After hardening

Figure 8.8   Lynis output before and after configuring banner message.

as the attackers will have system's OS and patch level information prior login to system. It will enable the attacker to carry out attacks related to the corresponding OS and patch. Hence there is a need to change the banner message. Permission */etc/issue./etc/issue.net* and */etc/motd* should be configured properly. The user id and group id should be 0/root, and access should be 644.

As per the U.S. Department of Defense guidelines, a warning message should have an organization name that owns the system. It should mention that the system user is trying to access is under monitoring, and the use of the system implies the user's consent to monitoring. Edit */etc/issue./etc/issue.net* file with the content as shown in below Figure 8.7. Figure 8.8 shows the lynis output before and after configuring banner message.

8.3.9 Configuring crons

Cron is considered a time-based job scheduler based in Unix systems used to run scripts or commands at fixed intervals. Tasks like maintenance and backups need to be performed regularly when no one uses the resource. System admin can execute the task at a specified time on a repetitive basis with the help of **Cron** and **at**. Even if users don't have any shell or jobs to execute at regular intervals, cron is still needed for the system to perform maintenance jobs like checking for updates or security monitoring.

Ensure correct permissions are set on */etc/cron.d, /etc/crontab, /etc/cron.weekly, /etc/cron.hourly, /etc/cron.daily, /etc/cron.monthly.* The **/etc/crontab** is a system-wide cron file. It contains information about system-level jobs run by cron. Read and write access can give the unprivileged user access to system-level information about cron tasks, providing the provision for privilege escalation. Hence Uid and Gid should be 0/root, and no permission should be given to the group and others, i.e., Access 600.

```
1  chown  root : root  /etc/crontab/
2  chown  og - rwx  /etc/crontab/
```

/etc/cron.hourly have all the cron tasks that will run on an hourly basis. System admin can change the file inside this directory by using a text editor. The only root user should be given read and write access (Access: 700). Regular should not have any access to this file.

```
1  chown  root : root  /etc/cron . hourly/
2  chown  og - rwx  /etc/cron . hourly/
```

/etc/cron.daily have all the cron tasks will run daily. Read or Write access to an unprivileged user on this directory can enable them to elevate their privilege. Hence Read/Write access to this directory should be restricted to root only(Access: 700).

```
1  chown  root : root  /etc/cron . daily/
2  chown  og - rwx  /etc/cron . daily/
```

/etc/cron.weekly have all the cron tasks that will run every week. Read or Write access to an unprivileged user on this directory can enable them to elevate their privilege. Hence Read/Write access to this directory should be restricted to root only(Access: 700).

```
1  chown  root : root  /etc/cron . weekly/
2  chown  og - rwx  /etc/cron . weekly/
```

/etc/cron.monthly have all the cron tasks that will run every month. Read or Write access to an unprivileged user on this directory can enable them to elevate their privilege. Hence Read/Write access to this directory should be restricted to root only(Access: 700).

```
1  chown  root : root  /etc/cron . monthly/
2  chown  og - rwx  /etc/cron . monthly/
```

/etc/cron.d is similar to directories as mentioned above, but it provides more granularity as to when cron jobs will run. Read/Write access should also be the same as directories mentioned above. A regular user should be refrained from reading and writing access to this directory as it might enable them to escalate privileges.

```
1  chown  root : root  /etc/cron . d/
2  chown  og - rwx  /etc/cron . d/
```

(a) Before hardening

(b) After hardening

Figure 8.9 Lynis output before and after setting permissions on cron.

Figure 8.9 shows the lynis output before and after setting permissions on cron.

8.3.10 User shell configuration

/etc/profile file is used to set system-wide environment variable and startup scripts. Initial values of PATH or PS1 are defined for all shell users of the systemetc/profile is executed only for an interactive shell. System admins can customize the system by making changes to this file. For large changes, the application-specific changes user needs to create a separate shell script(*.sh) inside */etc/profile.d/* directory. */etc/bash.bashrc* invoked for interactive as well non-interactive shell.

TMOUT environment variable sets a timeout of shell in seconds. It closes the shell of it inactive for n seconds. TMOUT variable restrains unauthorized users from accessing any other user's shell, which has been left open for a while.To configure *TMOUT* make the changes in */etc/profile, /etc/bash.bashrc* and place a bash file(e.g., tmout.sh) under */etc/profile.d/* directory. Run the following command to configure TMOUT. In the following command, the timeout is set as 900 seconds.

```
1 echo ''readonly TMOUT=900 ; export TMOUT''>>/etc/profile.d/
    tmout.sh
2 echo ''readonly TMOUT=900 ; export TMOUT''>>/etc/profile
3 echo ''readonly TMOUT=900 ; export TMOUT''>>/etc/bash.bashrc
```

UMASK stands for User Mask, also known as user file creation mask, which is used to assign default file permission when a new file or directory is created. In Linux machines, files are created with default permission of 666(rw-rw-rw-), and directories are created with default permission of

(a) Before hardening	(b) After hardening

Figure 8.10 Lynis output before and after setting UMASK value.

777(rwxrwxrwx). Umask value doesn't mean permission on files and directories. We need to subtract the umask value from the default permission of files and directories to calculate the actual permission. Default UMASK is 022, i.e., newly created files and directories will be readable by all users on the system. Here we are restricting the UMASK value as 027; hence permission for the newly created directory and files will be – *666-027 = 640*, i.e., files will be readable by users of the same Unix group. *777-027 = 750*, i.e., directories will be readable by the same Unix group users.

Run the following commands to set system-wide UMASK value:

```
1 echo ''umask 027'' >> /etc/profile.d/umaskval.sh
2 echo ''umask 027'' >> /etc/profile
3 echo ''umask 027'' >> /etc/bash.bashrc
```

Figure 8.10 shows the Lynis output before and after setting UMASK value.

8.3.11 USB devices

USBGuard is a software framework which helps in implementing whitelisting/blacklisting of USB devices, which protects the system from rogue USB devices. Run the below command to install USBGuard:

```
1 sudo apt-get install -y usbguard
```

/etc/usbguard/usbguard-daemon.conf is configuration file for USBGuard. This file is used by USBGuard daemon to load the policy rule set. We have made following changes to the attributes of *usbguard-daemon.conf*:

```
1 RestoreControllerDeviceState=false
2 PresentControllerPolicy=apply-policy
3 PresentDevicePolicy=apply-policy
4 InsertedDevicePolicy=apply-policy
5 ImplicitPolicyTarget=block
```

RestoreControllerDeviceState: The USBGuard daemon modifies some attributes of controllers, such as the default authorization status for new sub-device instances. With this setting, we can control whether the daemon will attempt to restore the attribute values to the state before modification on shutdown. On setting this value to true, the USB authorization policy can be bypassed by performing some daemon attack (via a local exploit or a USB device) to shut it down and restore the operating system's default state. Set this value to *false* to enhance security.

PresentControllerPolicy: How to handle USB controllers already connected when the daemon is started. One of the apply, reject, block, keep or apply-policy value will be assigned.

- *allow* - authorizes every device present on the system.
- *block* - deauthorize every device present on the system.
- *reject* - remove every device present on the system.
- *keep* - sync the internal state.
- *apply-policy* - ruleset is evaluated for every present device.

Setting value to apply-policy seems most secure, ensuring security even when the daemon hits a restart.

PresentDevicePolicy: How the devices are treated that is already connected when the daemon starts. It is setting the key value to apply policy.

InsertedDevicePolicy: How the USB devices are treated that are already connected after the daemon starts. Setting the key value to apply-policy.

ImplicitPolicyTarget: Policy that is to be applied to devices that don't match any rule in the policy.The key value is set to *block*.

To deauthorize USB devices on the system, use the following command.

```
1 for host in /sys/bus/usb/devices/usb*
2 do
3 echo 0 > $host/authorized_default
4 echo 0 > $host/authorized
5 done
```

Figure 8.11 shows the Lynis output before and after setting USBGuard.

8.3.12 Uncommon network protocol

Several uncommon network protocols are present on Linux kernels that are not commonly used. Hence it is recommended to disable those network protocols to reduce the attack surface of the system further. Some uncommon networks are Datagram Congestion Control Protocol (DCCP), Stream Control

(a) Before hardening

(b) After hardening

Figure 8.11   Lynis output before and after setting USBGuard.

Transmission Protocol (SCTP), Reliable Datagram Sockets (RDS), and Transparent Inter-Process Communication (TIPC). Network protocols can be disabled by creating or editing a file ending with .conf inside /etc/modprobe.d/ directory.

```
echo ''install dccp /bin/true''>> /etc/modprobe.d/
    uncommonprotocols.conf
echo ''install sctp /bin/true''>> /etc/modprobe.d/
    uncommonprotocols.conf
echo ''install rds /bin/true''>> /etc/modprobe.d/
    uncommonprotocols.conf
echo ''install tipc /bin/true''>>/etc/modprobe.d/
    uncommonprotocols.conf
```

Disable special-purpose services that are not required to reduce the system's attack surface further. The system admin should adequately review the open port. Each network port can be closed or opened by a firewall. Services listening on these ports sometime might lead to attacks if not checked. Hence unnecessary services should be removed from the system. Run the following and make sure all the services listed are required.

```
lsof -i -P -n | grep -v ''(ESTABLISHED)''
```

Remove unnecessary packages by using following command:

```
apt purge <package_name>
```

Some of them are Avahi Server, Domain Name System (DNS), Common Unix Print System (CUPS), Dynamic Host Configuration Protocol (DHCP), Network File System (NFS), File Transfer Protocol (FTP), HTTP server, IMAP and POP3 server, Samba, HTTP Proxy Server, Simple Network Management Protocol(SNMP) Server, rsync, Network Information Service (NIS), rsh client, talk, telnet, Lightweight Directory Access Protocol (LDAP), Remote Procedure Call (RPC).

```
apt purge avahi-daemon cups isc-dhcp-server nfs-kernel-
    server bind9 vsftpd apache2 dovecot-imapd dovecot-pop3d
    samba squid snmpd rsync nis rsh-client talk telnet ldap-
    utils rpcbind
```

8.3.13 Kernel hardening

To configure kernel parameter at runtime make the changes in */etc/sysctl.conf*. Following are the changes made to the parameters of */etc/sysctl.conf* file:

```
1  # Enable address space layout randomization (ASLR)
2  kernel.randarnize_va_space 2
3  # Restrict core dumps
4  fs.suid_dumpable 0
5  # Rejecting ICMP packets
6  net.ipv4.confall.accept_redirects 0
7  net.ipv4.confdefault.accept_redirects 0
8  net.ipv6.confall.accept_redirects 0
9  # Logging suspicious packets
10 net.ipv4.confall.log_martians 1
11 net.ipv4.confdefault.log_martians 1
12 # Rejecting source routed packets
13 net.ipv4.confall.accept_source_route 0
14 net.ipv4.confdefault.accept_source_route 0
```

Configure few more additional parameters (additonal step) to further increase the hardening index:

```
1  fs.protected_fifos 2
2  fs.protected_regular 2
3  Kernel.core_usespid 1
4  Kernel.dmesg_restrict 1
5  Kernel.kptr_restrict 2
6  Kernel.modules_disabled 1
7  Kernel.sysreg 0
8  Kernel.unprivileged_bpf disabled 1
9  net.core.bpf_jit_harden 2
```

Figure 8.12 shows the lynis Output before and after setting parameters in */etc/sysctl.conf*.

8.3.14 Compilers

Compiler or development tools are helpful for the user and any potential attackers. If there is an open compiler on the system, it becomes easier for

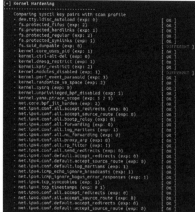

(a) Before hardening   (b) After hardening

Figure 8.12   Lynis output before and after setting parameters in */etc/sysctl.conf*.

attackers to run vulnerable programs. Some rootkits require the attacker to compile the program on the system. Hence, either disable the compilers or restrict them to any particular user or group. We have disabled the compilers by following commands:

```
1  chmod 000 /usr/bin/as>/dev/null
2  chmod 000 /usr/bin/byacc>/dev/null
3  chmod 000 /usr/bin/yacc>/dev/null
4  chmod 000 /usr/bin/bcc>/dev/null
5  chmod 000 /usr/bin/kgcc>/dev/null
6  chmod 000 /usr/bin/cc>/dev/null
7  chmod 000 /usr/bin/gcc>/dev/null
8  chmod 000 /usr/bin/*c++>/dev/null
9  chmod 000 /usr/bin/*g++>/dev/null
```

Sometimes there will be a symbolic link present for compilers. Changing just the permission might not work correctly. In that case, we need to remove the symbolic link. Table 8.1 shows the scenarios with hardening index and remarks.

8.3.15 Additional tools to enhance hardening index

AIDE (Advanced intrusion detection environment) is a tool used to check the integrity of files and directories, which helps detect an unauthorized change to the configuration files. A database of files is created by this utility, through

Table 8.1 Hardening index with scenario.

Scenario	Hardening index	Remarks
Fresh OS without update and upgrade	58	
Fresh Installation after update and upgrade	64	Updating and upgrading
After Using CIS benchmark	88	Following the steps provided in CIS benchmark
After following the suggestion from lynis and external resources	96	Disabling compilers, setting kernel parameters

which the integrity of files and folders is verified. AIDE can be installed on the Ubuntu system by using the following command:

```
1 apt install aide aide-common
```

To initialize AIDE run the following command:

```
1 aideinit
```

Clam AntiVirus is an open-source software to detect malicious software, trojans, viruses, and other malicious threats.

```
1 apt install clamav clamav-daemon
```

System Accounting (auditd) is a system administrator tool to monitor the system, such as system calls, file access, authentication failures, abnormal terminations, and program executions. The Linux Auditing system is not intended to provide protection. It provides awareness about the changes inside the system. Events are logged inside */var/log/audit/audit.log* file. There are limited rules defined by default. User needs to define their own rule when implementing auditd in */etc/audit/audit.rules* file.

```
1 apt install auditd audispd-plugins
```

fail2ban is an intrusion detection tool to protect the system from brute-force attacks. It bans IP addresses that show malicious activities like multiple password attempts to exploit the system. Fail2ban can be an updated firewall to block several malicious IP addresses.

```
1 apt-get install -y fail2ban
2 systemctl start fail2ban
3 systemctl enable fail2ban
```

Ansible

```
1 apt install -y software-properties-common
2 apt-add-repository-yes-update ppa:ansible/ansible
3 apt install -y ansible
```

8.4 Post Hardening

After performing all the hardening steps mentioned in the previous section, the lynis score of Ubuntu VM came out to be **96**. To install packages on the host machine user needs first to install them on VM. Based on the post-installation lynis score, the user can decide to install the package on the host machine. The user will know whether installing the package will compromise the system security or not, and based on that, and he can perform appropriate steps on the host machine. Package installation and score calculation script is placed inside VM (*getscore.sh*). We have customized the *apt* command with bash script and changed the environment variable to point *apt* command to custom script. On performing *apt* operation, the package will be first installed on the VM machine and based on the lynis score post-installation, the user will proceed on the host machine. Following are the steps of operation performed while installing the package :

```
1  apt install <package-name> on host machine.
2  Custom script for apt command will be called with parameter
       as package-name whose location is specified in
       environment variable.
3  Virtual Machine will be powered on.
4  Package will be installed on VM.
5  Lynis Score will be calculated post installation of package
       and returned to host.
6  There will be two cases :
7  a. Score will decrease means package has made changes to the
       system which have increased the attack surface of the
       system.
8  b. Score remains same means package have not affected the
       attack surface of system.
9  User will decide whether to install package on host or not.
10 VM is restored to previous snapshot.
```

8.5 Results

To get a list of all packages in the apt repositories first update is required, after which we can run the following command :

```
apt-cache search . |sort -d
```

The above command will also sort the package names as per the dictionary order. Approximately 70,000 packages are found in the fresh installation of Ubuntu18.04. *https://popcon.ubuntu.com/by_inst* ranks the Ubuntu packages as per the popularity. Some softwares cannot be directly installed via apt; we first need to add a repository and perform an apt install operation. Some softwares requires installation by *snap*. We have also tested the top 100 software in Ubuntu (e.g., Google Chrome, Dropbox, Sublime, Notepad, etc.)

Packages were tested using Python and bash script. Input file containing the list of packages has been given to the Python program. Bash script to install package and calculate the lynis score is placed inside the virtual machine. The virtual machine is started for each package in the input file, and the package is installed on the virtual machine. Post-installation, lynis score before and after installation is calculated on VM and returned to the host machine. These scores are logged inside a file on the host machine. VM is reset to the previous snapshot after each operation.

Note: We had to disable bootloader password to automate login through Virtualboxmanage command. Due to this Score has reduced to 95 from 96. Hence all the testing is done on lynis score of 95. Around 7000 and 450 top ranked packages were tested to get the following observation which is shown in Table 8.2. Figure 8.13 shows the hardening index of apt packages.

100 most popular apps like Chrome, Virtualbox, Dropbox, etc., gave the following results shown in Table 8.3 and Figure 8.14.

During a competition, there were 66 participants involved. Each one of them was assigned a playground for CTF, which was running on a docker container in the backend with *danielguerra/ubuntu-xrdp* image on it, which is based on the Ubuntu 18 server. Each participant can be considered an admin/user of an organization performing several tasks on their system. Hardening indexes for each participant were logged at regular intervals. Cron was set up on the master machine on which all the docker containers were running; using Python thread, we were able to run **lynis** simultaneously on all docker containers and stored the result to a log file. Default lynis score was 58. The following observation has been made shown in Table 8.4 and Figure 8.15.

Table 8.2 Hardening index post-installing Ubuntu official repository apt packages

Score	Number Of Packages	Remarks
88	2	Outdated packages or Unconfigured SSH
93	91	Compilers installed with default permission
94	27	Upgradable package is available and few packages are installing CUPS with default configuration
95	Remaining	No changes to hardening index

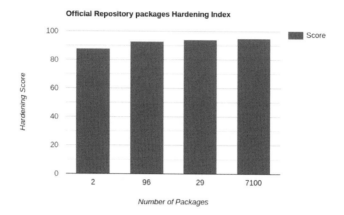

Figure 8.13   Hardening index of apt packages.

Table 8.3   Hardening index post-installing 100 popular Ubuntu software.

Score	Number of packages	Remarks
93	4	Compilers installed with default permission
94	2	Upgradable packages are available
95	Remaining	No changes to hardening index

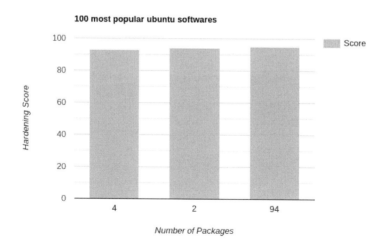

Figure 8.14   Hardening index of 100 popular Ubuntu softwares.

Table 8.4 Observations obtained for 66 participants.

Users	Remarks
46	Update and upgrade operation performed
5	Score increased to 61
3	Score decreased to 59 from 60
3	Score reached to 63
3	Score increased to 61 then back to 60
2	Score reached to 64
2	Score varied from 60-59-62-61, finally stable at 61
1	Score increased to 65 then back to 58
1	Increased to 63 then back to 60

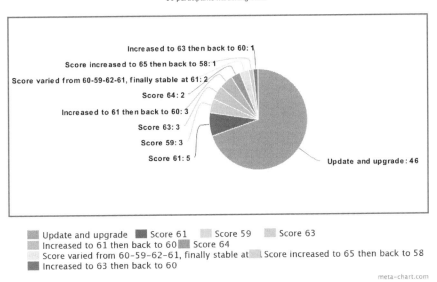

66 participants hardening index

Increased to 63 then back to 60: 1
Score increased to 65 then back to 58: 1
Score varied from 60-59-62-61, finally stable at 61: 2
Score 64: 2
Increased to 61 then back to 60: 3
Score 63: 3
Score 59: 3
Score 61: 5
Update and upgrade: 46

Update and upgrade Score 61 Score 59 Score 63
Increased to 61 then back to 60 Score 64
Score varied from 60-59-62-61, finally stable at Score increased to 65 then back to 58
Increased to 63 then back to 60

meta-chart.com

Figure 8.15 Hardening score of 66 participants.

8.6 Conclusion and Future Work

It is conclusive from the above results that installing packages even from authentic sources can make the system vulnerable to attacks. Packages often make configuration changes to the system, which does not comply with the security norms. Such changes need to be addressed immediately to protect the system from attacks. Apart from securing the network and firewalls, it is also essential to protect the system from the inside. System hardening helps reduce attack surface to a greater extent by closing most of the loopholes in the system. It is good practice to disable unused services and unused filesystem, enhance system authentication, keep a check on permissions, configure IP table rules and several other steps to reduce the attack surface.

Our work provides users with real-time identification of such security compromises. Moreover, as the packages are first installed on the sandboxed environment, the real system remains unchanged until the user installs it on the host machine based on the hardening index. Hence, the user is always aware of the particular package's changes to the system. Appropriate actions can be performed to enhance the security of the system. This work will be helpful for a system administrator for monitoring servers or hundreds of machines in an organization. We are currently getting a hardening score of 96, which can be attained to a maximum of 100. Our work is restricted to Ubuntu operating system currently. It can be further extended to other operating systems, like Windows and macOS. Automated installation is applicable only for packages available with *apt* package manager. Packages with *snap* and additional repositories need to be installed manually to VM. Such operation can also be automated.

Part IV

Honeypot

9

Stealpot Honeypot Network

Amardeep Singh[1], Om Prakash Mishra[2] and Sanjeev Kumar Sumbria

[1]eclerx services Ltd., India
[2]Canum infotech, India
E-mail: amardeepsg@gmail.com; om@canuminfotech.com;
sam288037@gmail.com

Abstract

A honeypot system is purposefully designed to be vulnerable and strategically deployed to be discovered as an easy target to attack. The objective is to continuously capture all possible data (source location, IP, approach, type, pattern, payload etc.) when an attempt or attack. The idea is to know more about different adopted attack methodologies and proactively build more specific cyber defense mechanisms. This work describes the basic concept and practical know-how on deploying a honeypot network using open-source platform components. The steps and elaborations have been presented, the way it was implemented as part of actual project work. Cybersecurity enthusiasts can apply this to explore easy deployment options for the honeypot network as a defensive security mechanism.

9.1 Introduction

With the rapid increase of digitization, cybersecurity has become more critical than ever. As it happens with everything in the limelight, the bad actors are yet more active now. Increasing reliance on digital means a bigger pie to exploit, newer tactics, techniques, and payloads continuously deployed by the threat actors. Security teams have always been catching up, running scheduled scans, publishing new vulnerabilities, and ensuring an on-time response. However, more often than not, these approaches do not ensure total safety.

225

We need more relevant and to the point threat intelligence inputs by trapping specific data and methods being adopted by possible threats in action (IOCs & TTPs). It would allow security teams to prioritize defense against specific threats in real-time instead of running behind the target of keeping everything up to date and still getting impacted by an attack. Honeypots help with such threat intelligence inputs. Honeypots have been around for more than two decades but are probably considered expensive and complex to manage and maintain. It requires automated setup, easily configurable, orchestrated deployment mechanisms, and central management at affordable costs. The Stealpot sources its evolution based upon these ideas and asks.

We are going through challenging times, and things are changing at the speed of light. It is no longer feasible to have tall perimeter walls (Build perimeter security by deploying NGFWs, IPS/IDS, DDOS, etc.) and be assured of total security within the boundaries. The idea of a shield around and staying protected from the threats around could just be imagined. We have cyber threats everywhere. Data is distributed across the boundaries of any defined periphery. There are increasing trends of multi-cloud, hybrid, on-premise and all possible combinations of deployments. End-users are distributed on various networks with more BYOD (Bring your own device) and work from home requirements. In such scenarios, how do we spell the relevance of perimeter security? Can we define perimeters? Do they exist?

Given the complexity and dynamics, zero trust could give us some answers. With a rapidly evolving threat landscape, honeypots have gained relevance and can be leveraged effectively if they are managed well with automated, orchestrated and continuous deployment mechanisms. Of course, ease is an ask for both speed and scale to realize actual yield. Here, we present an approach to creating a honeypot ecosystem comprising a central reporting and management console, remote honeypot deployment and integration with a malware analysis engine using various open-source platforms.

Honeypot(s) are mainly used by defensive security product/solutions companies. These companies deploy high-interaction research honeypots across the globe to gather intelligence, capture intelligence like origin, preferred targets, techniques, motivation, etc., to feed into their solutions as signatures, patterns, behaviors, etc. Industry at large is not using honeypots, instead of relying on products developed by security vendors using this intelligence. Figure 9.1 is talking about the usage trend of honeypots as we see it today. These were taken from a survey conducted under the supervision of SANS Institute [22].

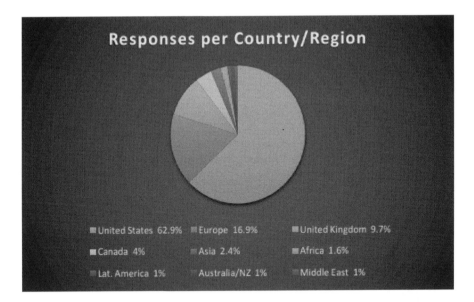

Figure 9.1 Responses per country/region.

The hypothesis that honey technologies are not widely deployed in production systems proved correct. Of those surveyed, only 25% had honeypots. Another aspect worth looking at was the breakdown of the responding organizations. Interestingly, the top three organization types were privately-held corporations, publicly-traded corporations, and government agencies. Academic institutions, utilities, and non-profits also made an appearance. It is worth noting that 34.6% of respondents did not identify the type of organization, however, so this data point might not offer extraordinary insights. Figure 9.2 shows the organization type breakdown by percentage.

One use case that is observed in the industry is to replicate a critical production application as a honeypot. It is expensive to deploy and maintain and only attempt to protect the specific application, not something everyone can attempt to use [64].

We suggest a slightly different approach, deploying a group of low-interaction honeypots managed through a central console that provides reporting and analytics. It is not expensive, less complex, still provides intelligence on prevailing threats closer to where you are present as an organization, and protects your overall IT footprint and not specific applications. Most importantly, we attempt to do this using open-source platforms/technologies and use an architecture that can be easily modified and extended to match

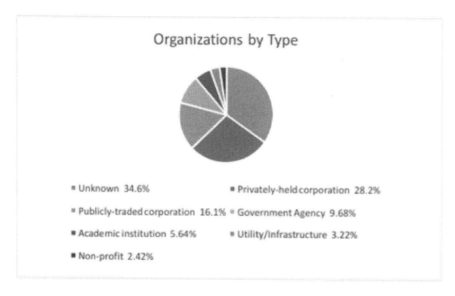

Figure 9.2 Organizations by type.

an organization's specific requirement using the same main components (for central management, reporting, and analytics).

We call our honeypot "Stealpot." Stealpot, a honeypot network, as the name suggests, is a network of different kinds of honeypots deployed across geographic locations to cajole prevalent threat actors into attempting a hack and help collection of IOCs, payloads, tactics and techniques. It is a centrally managed honeypot network of different honeypots that can help security teams identify relevant threats and fine-tune their defenses proactively.

9.1.1 Problem statement

Composing a high-interaction honeypot network [67] with the following key features:

- Easy to configure and deploy
- Central management
- Cost-effective implementation
- Extendible analytics and reporting capabilities

Considering the above key requirements, the Stealpot solution starts with an approach to leverage open-source. We experimented with many open-source platforms and components. With due research and learnings, we preferred

MHN (Modern honeypot network) [25], which has a solid central management platform for honeypots. However, the support community has not kept up to date for MHN.

9.2 Methodology

Following are the ingredients of our dish:

- DigitalOcean developer cloud platform. (Any cloud platform would be fine)
- MHN (Modern honeypot network): Open-source central honeypot deployment and reporting platform. Figure 9.3 shows the MHN console.
- Honeypots to cover multiple ports and protocols for collection of IOCs -
  - Dionaea
  - Cowrie
  - Elastichoney
  - Shockpot
- **ELK Stack** – Elasticsearch, Logstash, and Kibana [48]: For expandable analytics, reporting and integration with honeypots that do not have an out of the box support with MHN.
- **VirusTotal API script** – To analyze payloads uploaded to Dionaea for malicious content and inject details into Elasticsearch and Kibana for analytics and reporting.

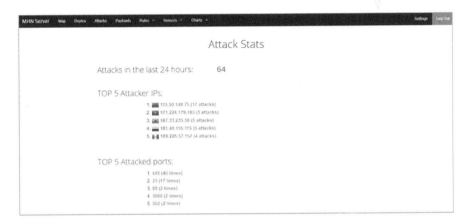

Figure 9.3   MHN console.

We focused on creating a relatively high-interaction honeypot network without replicating an entire production setup. So, we picked up MHN as the central component deployed on the cloud. We added four different honeypots in the network, each capturing the IOCs for different protocols. To keep the honeypots relatively more vulnerable, we utilized Ubuntu 18.04 as the OS platform for all honeypots and other components in the network.

Cloud deployment was an obvious choice for implementation speed and easy capturing of IOCs across different geographies. It can be deployed in separate VLAN(s) in the enterprise network. A compromised honeypot does not allow any compromise beyond the VLAN(s) with the honeypot.

9.3 Architecture: Keeping It Simple and Straightforward

Figure 9.4 shows the overall architecture of the Stealpot.

9.3.1 Components

Platform: MHN (Modern honey network) deployed on an Ubuntu 18.04 VM hosted on a cloud platform. Figure 9.3 shows the MHN console.

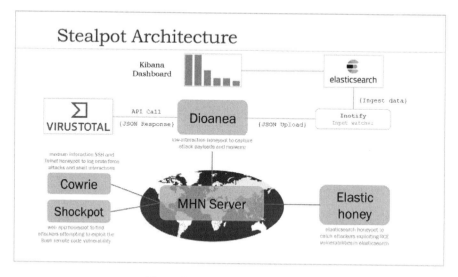

Figure 9.4 Architecture diagram.

9.3.2 Honeypots: Ubuntu 18.04 as the base OS, deployed at four different regions across the globe

- **Dionaea** – For capturing the most prevalent malware samples and other IOCs.
- **Cowrie** – For identifying SSH and Telnet attack statistics.
- **Elastichoney** – For capturing specific threats to ELK stack as this forms part of most big data deployments.
- **Shockpot** – Webapp Honeypot to capture statistics for Bash RCE vulnerability. (Do people still exploit it?)

9.3.3 Other components/integration

- **Dionaea integration with VirusTotal:** To capture analysis data and upload it to an ELK stack for analytics and dashboarding purposes.
- **"inotifywait":** Custom script for monitoring a folder and triggering file upload to VirusTotal as soon as a suspected file gets uploaded via FTP. Figure 9.5 shows the corresponding script. It was hosted as part of the Dionaea honeypot.

```
inotifywait -m -e create -e moved_to --format "%f" $TARGET \
        | while read FILENAME
                do
                        sleep 2s
                        python3 vt_upload_cmd_line.py $TARGET$FILENAME
                done
```

Figure 9.5 Snippet of the Trigger script for uploading to VirusTotal.

```
#Put the Virustotal API Key in the header
headers = {'x-apikey': '<API KEY>'}

#taking file from command line argument. For example: python3 vt_upload.py path/to/file/wannacry.exe
file_from_hp = sys.argv[1]

#Function to check the ID of the file that has been uploaded
def check_id_of_uploaded_file(id_of_file,hash_of_current_file):
        os.system("sleep 300")

        #URL of the file concat with the ID
        url_get_file = 'https://www.virustotal.com/api/v3/analyses/'+id_of_file

        #get request to get the result from virustotal and convert it into json
        response_file_upload_virustotal = requests.get(url_get_file, headers=headers).json()

        #Beautifying the JSON.
        print(json.dumps(response_file_upload_virustotal, sort_keys=False, indent=4))

        with open("/output_json/"+hash_of_current_file+".json", "w") as outfile:
                json.dump(response_file_upload_virustotal, outfile, indent=4)
```

Figure 9.6 Code for making API call to VirusTotal.

Figure 9.6 shows the snippet of the Python script used for making API call to VirusTotal, submitting a suspect file for analysis and receiving analysis in a json format output file [53]. See Elasticsearch and Kibana in action, Figure 9.7 shows a sample command to upload required details from VirusTotal json analysis file into Elasticsearch. Figures 9.8–9.13 depict the tool in action and showcase the IOCs captured in a very short period of time.

```
curl -X PUT "http://hostname:9200/<index name>/_settings?pretty" -H 'Content-Type: application/json'
-d' { "index.mapping.total_fields.limit": 2000 }'
```

Figure 9.7 Curl command to upload content from VT output file on Dionaea in json format into the Elasticsearch for indexing, analysis, and dashboard.

	Name	Hostname	IP	Honeypot	UUID
1-	hp41-dionaea	hp41	143.198.103.66	dionaea	3147a398-eeb0-11eb-b0de-3ae5a8cfbe3d
2-	HPC-cowrie	HPC	206.189.21.51	cowrie	95d63e50-eebd-11eb-b0de-3ae5a8cfbe3d
3-	HPEH-elastichoney	HPEH	104.248.151.121	elastichoney	a759b906-eebe-11eb-b0de-3ae5a8cfbe3d
4-	HPp0-shockpot	HPp0	167.99.252.214	shockpot	90a65b30-eecf-11eb-b0de-3ae5a8cfbe3d

Figure 9.8 IOCs captured within minutes of powering on.

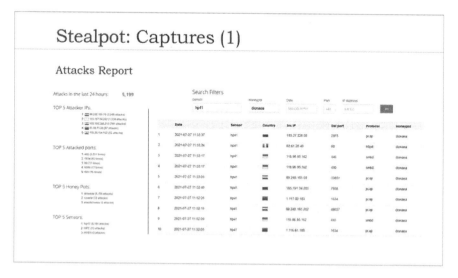

Figure 9.9 Attack report.

Search filters

Sensor	Honeypot	Date	Port	IP Address	
HPC	**cowrie**	**MM-DD-YYYY**	**445**	**8.8.8.8**	**GO**

S.No	Date	Sensor	Country	Src IP	Dst Port	Protocol	Honeypot
1	2021-07-27 11:39:09	HPC	Poland	193.169.254.138	22	ssh	cowrie
2	2021-07-27 11:28:50	HPC	US	8.21.11.108	22	ssh	cowrie
3	2021-07-27 11:24:05	HPC	Vietnam	103.226.250.169	22	ssh	cowrie
4	2021-07-27 11:23:48	HPC	Vietnam	103.226.250.169	22	ssh	cowrie
5	2021-07-27 11:22:15	HPC	Singapore	188.166.247.82	22	ssh	cowrie
6	2021-07-27 11:20:17	HPC	South Africa	41.193.25.8	22	ssh	cowrie
7	2021-07-27 10:45:16	HPC	US	128.14.141.42	22	ssh	cowrie
8	2021-07-27 10:00:08	HPC	Lithuania	141.98.10.203	22	ssh	cowrie
9	2021-07-27 09:59:57	HPC	Lithuania	141.98.10.203	22	ssh	cowrie
10	2021-07-27 09:59:46	HPC	Lithuania	141.98.10.203	22	ssh	cowrie

Search filters

Sensor	Honeypot	Date	Port	IP Address	
HPEH	**elastichoney**	**MM-DD-YYYY**	**445**	**8.8.8.8**	**GO**

S.No	Date	Sensor	Country	Src IP	Dst Port	Protocol	Honeypot
1	2021-07-27 10:50:30	HPEH	US	192.241.221.242	9200	http	elastichoney
2	2021-07-27 10:50:08	HPEH	US	192.241.215.196	9200	http	elastichoney

Figure 9.10   Attack report.

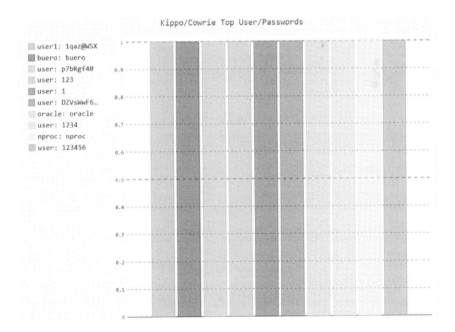

Figure 9.11   Sample IOCs for malware uploaded on Dionaea.

daddr	saddr	dport	sport	sha512
143.198.103.66	200.75.29.82	445	62370	de74b133db3267c323dd84f85915fa6df37149386016315f47f33c284d2d1669a75ac64a2633f267787dd04c2d1f8ee61f381ac6b1ccfd81a1c7de3cd80814aafd
143.198.103.66	208.107.144.88	445	49441	308246d2a09526d0c6536e5d5ef3ae9a1f11373391926fbeb20f4cea973853314a33e4a1228144e0ec2002bd6c3f57eb00b4ec1b262e9dab445688d09a4ace52c
143.198.103.66	31.135.205.10	445	49492	1c0958149874900419b9da2431ea57dd3f2a36f7d441305c36205fe150bb6c03677e389f4d231f326085288c4c0bd8c9e46d997b058a2019fac013ed1120fa75
143.198.103.66	106.195.1.249	445	55810	4ee77f041a92a369e834f7cd6be84f57a874cd534c89b1c473391f616d43a66080634fc8aabfa9dacb2e665a862b0c300bee9b67211f52e1ec243ddc647f6f1f66
143.198.103.66	31.173.18.41	445	63652	7d49822b547da91003139c18b0384c047e728540dfd49e0f7e4284d1ed1ef941e03d464af7051edc7d12cfe3dfbeb28b5acd40dcaa8fd6d4175dde5222871fa6
143.198.103.66	180.183.123.101	445	51289	4ea3837e9c852575fba17aac140791e1ecd41d1316e7d16c9b95f3a10ed70d244f03df6533440ab24571dd5439d6ad5eb1446d4c99858c7a22cbe672a3be6b97
143.198.103.66	77.222.121.109	445	59450	eaa97fca758267a98cb5bf2fb6.19c3507212dbd519677db9069043e2b7103c4d7f407d4de64f805.3b7e5d82567279871795606bd037825c68f55794840de5033f
143.198.103.66	179.35.211.178	445	61941	9464ab88ccb6adce94d95570b0c902f384194a736cfc6aeb9a47b9dd150572e6abce60c1d362dbe739684faff25a5a0829ca5ecbb455c4647na7f3ad5b5d693
143.198.103.66	46.182.132.28	445	61399	6cfa63ad1be4817ae7c0c363db4336f069nde66e52e7aa0efc5ff5b72c58b3350d6982o790b79455f2430f9e394f70421130f62a5b22a89818e34fd518812e112d
143.198.103.66	58.56.153.2	445	61010	b7f39511cbad53fec9d80dd78b02def424d6ded3a77085757349ae4ee1b713656712ea81a32789eb7899d9d35fbe72aae1bcbee44ff427bca59043eb838ea29

Figure 9.12   Malware uploaded to Dionaea.

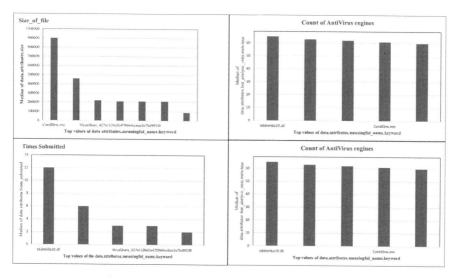

Figure 9.13   Malware visualization.

9.4 Conclusion

Honeypots can become an integral part of a defensive strategy to identify the real threats, both internal and external and feed that intelligence into your defensive security platforms in the form of IOCs and TTPs. The biggest hindrance in achieving the same is complexity and cost. Our approach uses open source technologies to create a comprehensive honeypot network, keeping the architecture flexible and straightforward.

We solve the problem of complexity, cost, and adaptability by focusing on a ready-to-use out-of-the-box central console, ready to deploy low-interaction honeypots that can be deployed using MHN. The central management and deployment solution and ELK stack for adding flexibility and extendibility to the solution. We hope to increase the usage of honeypot and honeypot networks in the corporate world to protect against internal and external threats by offering a simple, low-cost solution. At the end of the day, the more we know about our enemies, the better we can protect ourselves against them.

References

[1] A06_2021vulnerable_and_outdated_components. https://owasp.org/To
p10/A06_2021-Vulnerable_and_Outdated_Components/, 2021.

[2] 2021 open source security and risk analysis report. https://www.synops
ys.com/software-integrity/resources/analyst-reports/open-source-secu
rity-risk-analysis.html, 2022.

[3] Flowchart maker and online diagram software. https://app.diagrams.net/,
2022.

[4] Kotlin v 1.6.10. https://kotlinlang.org/, 2022.

[5] Owasp dependency-check. https://owasp.org/www-project-dependency
-check/, 2022.

[6] Mamoun Alazab and Sitalakshmi Venkatraman. Detecting malicious
behaviour using supervised learning algorithms of the function calls.
International Journal of Electronic Security and Digital Forensics,
5(2):90–109, 2013.

[7] Mamoun Alazab, Sitalakshmi Venkatraman, Paul Watters, Moutaz
Alazab, et al. Zero-day malware detection based on supervised learning
algorithms of api call signatures. 2010.

[8] Inc. Amazon Web Services. Open distro. https://opendistro.github.io/for
-elasticsearch/, 2021.

[9] Saptarshi Bej, Narek Davtyan, Markus Wolfien, Mariam Nassar, and Olaf
Wolkenhauer. Loras: An oversampling approach for imbalanced datasets.
Machine Learning, 110(2):279–301, 2021.

[10] BlueVirtualNerds-Team. Bluevirtualnerds/g0rking. https://github.com/B
lueVirtualNerds/G0rKing/blob/main/DorkingGuide, 2021.

[11] L. Breiman, J. H. Friedman, R. A. Olshen, and C. J. Stone. *Classification
and Regression Trees*. Wadsworth and Brooks, Monterey, CA, 1984.

[12] Leo Breiman. Random forests. *Mach. Learn.*, 45(1):5–32, October 2001.

[13] Ferhat Ozgur Catak, Ahmet Faruk Yazı, Ogerta Elezaj, and Javed
Ahmed. Deep learning based sequential model for malware analysis
using windows exe api calls. *PeerJ Computer Science*, 6:e285, July 2020.

[14] Danqi Chen and Christopher Manning. A fast and accurate dependency parser using neural networks. In *Proceedings of the 2014 Conference on Empirical Methods in Natural Language Processing (EMNLP)*, pages 740–750, Doha, Qatar, October 2014. Association for Computational Linguistics.

[15] B. Cheng, Q. Tong, J. Wang, and W. Tian. Malware clustering using family dependency graph. *IEEE Access*, 7:72267–72272, 2019.

[16] Aviad Cohen, Nir Nissim, and Yuval Elovici. Maljpeg: Machine learning based solution for the detection of malicious jpeg images. *IEEE Access*, 8:19997–20011, 2020.

[17] Aviad Cohen, Nir Nissim, and Yuval Elovici. Maljpeg: Machine learning based solution for the detection of malicious jpeg images. *IEEE Access*, 8:19997–20011, 2020.

[18] KHT Dam and T Touili. Malware detection based on graph classification. *ICISSP*, pages 455–463, 2017.

[19] Jeremie Daniel. Install and setup cowrie honeypot on ubuntu (linux). https://medium.com/@jeremiedaniel48/install-and-setup-cowrie-hone ypot-on-ubuntu-linux-5d64552c31dc, 2019.

[20] DigitalOcean. How to create a droplet from the digitalocean control panel. https://docs.digitalocean.com/products/droplets/how-to/create/, 2021.

[21] Wazuh Docs. User manual. https://documentation.wazuh.com/current/us er-manual/index.html, 2021.

[22] Andrea Dominguez. The state of honeypots: Understanding the use of honey technologies today. https://www.sans.org/white-papers/38165/, 2022.

[23] Gintare Karolina Dziugaite, Zoubin Ghahramani, and Daniel M Roy. A study of the effect of jpg compression on adversarial images. *arXiv preprint arXiv:1608.00853*, 2016.

[24] Phibos et al. Dinotools or dionaea. https://github.com/DinoTools/diona ea/graphs/contributors, 2022.

[25] Serpulga et al. Modern honey network. https://github.com/pwnlandia/m hn, 2022.

[26] Firehol. All cybercrime ip feeds. http://iplists.firehol.org/, 2022.

[27] David French. How to setup "cowrie" – an ssh honeypot. https://medium .com/threatpunter/how-to-setup-cowrie-an-ssh-honeypot-535a68832e 4c, 2018.

[28] Yoav Freund and Robert E. Schapire. A decision-theoretic gener- alization of on-line learning and an application to boosting. In *Proceedings of the Second European Conference on Computational*

Learning Theory, EuroCOLT '95, page 23–37, Berlin, Heidelberg, 1995. Springer-Verlag.

[29] Andrea Galassi, Marco Lippi, and Paolo Torroni. Attention in natural language processing. *IEEE Transactions on Neural Networks and Learning Systems*, page 1–18, 2020.

[30] Ian J Goodfellow, Jonathon Shlens, and Christian Szegedy. Explaining and harnessing adversarial examples. *arXiv preprint arXiv:1412.6572*, 2014.

[31] Claudio Guarnieri. Cuckoo Sandbox - Automated Malware Analysis. https://cuckoosandbox.org/, 2020.

[32] Common Vulnerability Scoring System V3.1: User Guide. Cve overview. https://www.first.org/cvss/user-guide, 2021.

[33] Eric Hamilton. Jpeg file interchange format. 2004.

[34] Dan Hendrycks and Kevin Gimpel. Early methods for detecting adversarial images. *arXiv preprint arXiv:1608.00530*, 2016.

[35] Sepp Hochreiter and Jürgen Schmidhuber. Long short-term memory. *Neural Comput.*, 9(8):1735–1780, November 1997.

[36] DigitalOcean Inc. Digital ocean. https://www.digitalocean.com/, 2022.

[37] H. Jiang, T. Turki, and J. T. L. Wang. Dlgraph: Malware detection using deep learning and graph embedding. In *2018 17th IEEE International Conference on Machine Learning and Applications (ICMLA)*, pages 1029–1033, 2018.

[38] Yasoob Khalid. Understanding and decoding a jpeg image using python. https://yasoob.me/posts/understanding-and-writing-jpeg-decoder-in-python/, 2022.

[39] Ban Mohammed Khammas. The performance of iot malware detection technique using feature selection and feature reduction in fog layer. In *IOP Conference Series: Materials Science and Engineering*, volume 928, page 022047. IOP Publishing, 2020.

[40] Chan Woo Kim. Ntmaldetect: A machine learning approach to malware detection using native api system calls. 02 2018.

[41] Konstantinos Kosmidis and Christos Kalloniatis. Machine learning and images for malware detection and classification. 09 2017.

[42] ktsaou. firehol blocklist ipsets. https://github.com/firehol/blocklist-ipsets/blob/master/README.md#list-of-ipsets-included, 2022.

[43] Sushil Kumar et al. An emerging threat fileless malware: a survey and research challenges. *Cybersecurity*, 3(1):1–12, 2020.

[44] Rakesh Singh Kunwar and Priyanka Sharma. Framework to detect malicious codes embedded with jpeg images over social networking

sites. In *2017 International Conference on Innovations in Information, Embedded and Communication Systems (ICIIECS)*, pages 1–4. IEEE, 2017.

[45] Alexey Kurakin, Ian Goodfellow, Samy Bengio, et al. Adversarial examples in the physical world, 2016.

[46] Chatchai Liangboonprakong and Ohm Sornil. Classification of malware families based on n-grams sequential pattern features. In *2013 IEEE 8th Conference on Industrial Electronics and Applications (ICIEA)*, pages 777–782. IEEE, 2013.

[47] Ching-Yung Lin and Shih-Fu Chang. A robust image authentication method distinguishing jpeg compression from malicious manipulation. *IEEE Transactions on Circuits and Systems for Video Technology*, 11(2):153–168, 2001.

[48] Rose Web Services LLC. Install elasticsearch, logstash, and kibana on ubuntu 20.04. https://www.rosehosting.com/blog/how-to-install-elk-stac k-on-ubuntu-20-04/, 2022.

[49] Johnny Long. j0hnnyhax. https://en.wikipedia.org/wiki/Johnny_Long, 2022.

[50] Luca Massarelli, Giuseppe Luna, Fabio Petroni, and Leonardo Querzoni. Safe: Self-attentive function embeddings for binary similarity. 11 2018.

[51] Microsoft. Visual studio v 17.1. https://kotlinlang.org/, 2022.

[52] Tomas Mikolov, Kai Chen, Greg Corrado, and Jeffrey Dean. Efficient estimation of word representations in vector space, 2013.

[53] Subhasis Mukhopadhyay. Python implementation of virustotal vt3 api and automation. https://github.com/SubhasisMukh/virustotal-vt3, 2020.

[54] K Nasla and M Shabna. Enhanced maljpeg: A novel approachfor the detection of malicious jpeg images. 2020.

[55] L. Nataraj. A signal processing approach to malware analysis. 2015.

[56] Blue Virtual Nerds. Blue virtual nerds. sauron.in, 2022.

[57] NGINX. Configuring https servers. http://nginx.org/en/docs/http/config uring_https_servers.html, 2022.

[58] US Department of Homeland Security. About the cve program. https: //www.cve.org/About/Overview, 2021.

[59] US Department of Homeland Security. Common vulnerabilities and exposures. https://cve.mitre.org/, 2021.

[60] Angelo Oliveira. Malware analysis datasets: Api call sequences, 2019.

[61] Morten Oscar Østbye. Multinomial malware classification based on call graphs. 2017.

[62] OWASP. Owasp g0rking. https://owasp.org/www-project-g0rking/, 2021.

[63] Nicolas Papernot, Patrick McDaniel, Ian Goodfellow, Somesh Jha, Z Berkay Celik, and Ananthram Swami. Practical black-box attacks against machine learning. In *Proceedings of the 2017 ACM on Asia conference on computer and communications security*, pages 506–519, 2017.

[64] Reshma R Patel and Chirag S Thaker. Zero-day attack signatures detection using honeypot. In *International Conference on Computer Communication and Networks (CSI-COMNET)*, 2011.

[65] Abdurrahman Pektaş and Tankut Acarman. Deep learning for effective android malware detection using api call graph embeddings. In *Soft Computing*, pages 1027–1043, 2020.

[66] PhoenixNAP. Setup and use nginx as a reverse proxy. https://phoenixnap.com/kb/nginx-reverse-proxy, 2021.

[67] Honeynet Project. The honeynet project. https://www.honeynet.org/, 2022.

[68] Rapid. How to install suricata nids on ubuntu. http://nginx.org/en/docs/http/configuring_https_servers.html, 2022.

[69] Rapid7 Report. Malware attacks: Definition and best practices. https://www.rapid7.com/fundamentals/malware-attacks/, 2022.

[70] SankalpIT. How to create slack incoming webhook url. https://sankalpit.com/plugins/documentation/how-to-create-slack-incoming-webhook-url/, 2022.

[71] Slack. Sending messages using incoming webhooks. https://api.slack.com/messaging/webhooks, 2022.

[72] Suricata. Rule management with suricata-update. https://suricata.readthedocs.io/en/suricata-6.0.0/rule-management/suricata-update.html, 2019.

[73] Suricata. Setting up ips/inline for linux. https://suricata.readthedocs.io/en/suricata-6.0.0/setting-up-ipsinline-for-linux.html, 2021.

[74] Mingdong Tang and Quan Qian. Dynamic api call sequence visualisation for malware classification. *IET Information Security*, 13(4):367–377, 2018.

[75] Ashish Vaswani, Noam Shazeer, Niki Parmar, Jakob Uszkoreit, Llion Jones, Aidan N. Gomez, Lukasz Kaiser, and Illia Polosukhin. Attention is all you need, 2017.

[76] Duc-Ly Vu. Deepmal: Deep convolutional and recurrent neural networks for malware classification. 10 2020.

[77] Juntao Wang and Xiaolong Su. An improved k-means clustering algorithm. In *2011 IEEE 3rd International Conference on Communication Software and Networks*, pages 44–46, 2011.

[78] Wazuh. Ruleset. https://documentation.wazuh.com/current/user-manual/ruleset/, 2022.

[79] Wazuh. Using osint to create cdb lists and block malicious ips. https://wazuh.com/blog/using-osint-to-create-cdb-lists/, 2022.

[80] Wikepedia. Elf. https://en.wikipedia.org/wiki/Elf, 2022.

[81] Wikepedia. Malware. https://en.wikipedia.org/wiki/Malware, 2022.

[82] Fei Xiao, Zhaowen Lin, Yi Sun, and Yan Ma. Malware detection based on deep learning of behavior graphs. *Mathematical Problems in Engineering*, 2019, 2019.

[83] Lu Xiaofeng, Zhou Xiao, Jiang Fangshuo, Yi Shengwei, and Sha Jing. Assca: Api based sequence and statistics features combined malware detection architecture. *Procedia Computer Science*, 129:248–256, 2018.

[84] Kai Zhang, Chao Li, Y. Wang, Xiaobin Zhu, and H. Wang. Collaborative support vector machine for malware detection. In *ICCS*, 2017.

Index

About the Editors

Anand Handa – is a senior research engineer and a post-doctoral fellow at C3i center, IIT Kanpur. His focus areas include malware analysis, memory forensics, IDS, etc. His role at C3i involves working on projects with malware analysis and IDS as a significant component. He has published his work at various international conferences and journals of repute. He is an active member of different working groups.

Rohit Negi – An IIT Kanpur alumni with over 10 years of experience specialising in industrial automation. A well-rounded researcher with a background in areas related to cyber defense of ICS and OT layer. Actively involved in development and incubation of indigenous solutions that improve cyber defense strategies and capabilities, such as ICS-honeypots, ICS-SIEM, Threat intelligence, Anomaly based IDS, etc.Lead engineer, security architect and security operations lead at C3i Center, IIT Kanpur. Published several international research papers and written several book chapters.

S. Venkatesan – is an associate professor at the Department of Information Technology at the Indian Institute of Information Technology Allahabad (IIITA). He heads IIITA's C3iHub IoT Security Lab and is a member of the Network Security and Cryptography (NSC) Group. He has authored several research papers published in reputed journals and presented at conferences. His research interests include network security, cloud computing, social network privacy, mobile agent security, applied cryptography, and blockchain.

Sandeep K. Shukla – is a professor of computer science and engineering with the Indian Institute of Technology. He is an IEEE Fellow, ACM distinguished scientist, and subject matter expert in cybersecurity of cyber-physical systems and blockchain technology. He is a recipient of various prestigious honors, and he serves as a joint coordinator for the C3i Center and the National Blockchain Project at IIT Kanpur, India.